Environmental Regulation and Rationality

Multidisciplinary Perspectives

Environmental Regulation and Rationality

Multidisciplinary Perspectives

Edited by
Suzanne C. Beckmann and
Erik Kloppenborg Madsen

AARHUS UNIVERSITETSFORLAG

Environmental Regulation and Rationality
Multidisciplinary Perspectives

Edited by Suzanne C. Beckmann
and Erik Kloppenborg Madsen
Cover by Kitte Fennestad
Printed in Denmark by
Narayana Press, Gylling

ISBN 87 7288 912 8

Published with financial support from
The Danish Environmental Research Programme

AARHUS UNIVERSITY PRESS

Langelandsgade 177
DK-8200 Aarhus N
Fax (+45) 89 42 53 80
www.unipress.dk

73 Lime Walk
Headington, Oxford OX3 7AD
Fax (+44) 1865 750 079

Box 511
Oakville, CT 06779
Fax (+1) 860 945 9468

CONTENTS

INTRODUCTION

Environmental Regulation and Rationality

Erik Kloppenborg Madsen and Suzanne C. Beckmann

I

THIS COLLECTION OF ESSAYS has rationality in behaviour and in behavioural regulation as its central theme. More specifically, these articles address the need for behavioural change called for by the environmental concern voiced in popular debate, and they reflect an ongoing effort to discover what it will take to improve the environmental behaviour of people, households, and commercial concerns. The essays also address the issue of what the concept of regulation entails, and whose concern it could or should be – a top-down, or a bottom-up process? Should regulation of behaviour be seen as the concern of public authorities, experts, or other actors involved with steering the actions of ordinary citizens? Or rather, should regulation – as suggested by Sagoff (1988) – be understood in the spirit of collective societal self-improvement, thus focusing on improving lifestyles and enhancing democratic and participatory social institutions?

The idea of writing this book was born in an interdisciplinary group of people working with social science research on environmental regulation since the early 1990's*. Soon after CeSaM was established, it

* The Centre for Social Science Research on the Environment (CeSaM) (www.au.dk/cesam); financially supported by The Danish Environmental Research Programme, and comprising some fifty researchers at nine institutions in Denmark.

became evident that the various members of the group emphasise different perspectives and different assumptions concerning the motivation and rationality of human actors, mainly because they represent different disciplines. Rationality thus became one of the central concepts around which our discourse in the group evolved.

A sincere interest in gaining an understanding of the core ideas of other disciplines developed in the course of interdisciplinary meetings, further motivating this endeavour to outline our individual disciplinary concepts of rationality. Discussion of the topic was organised in the following dimensions: assumptions about the actors, assumptions about structures and institutions, assumptions about the interplay between actors(s) and institutions, assumptions about nature and the nature/culture divide, and, finally, the specific use of the concept of rationality. The disciplines represented in the group are philosophy, theology, ethics, law, jurisprudence, political science, economics, marketing, and economic psychology.

It might be argued that trying to understand the ideas of other disciplines is in itself a rational enterprise, since the concept of rationality is understood both differently and similarly across and within disciplines. Although there are different views on what rationality is all about, these differences do not necessarily coincide with distinctive disciplines. Different concepts of rationality may be debated within a single discipline, but different disciplines may also share identical conceptions of rationality. Furthermore, the concept of rationality does not seem to play the same central role in the different disciplines. In economics, the *homo economicus* metaphor is a core assumption concerning the motivational and cognitive makeup of the actor. In jurisprudence, there seems to be less interest in assumptions as to actors' reasons for behaving in a certain way. Norms, on the other hand, are obviously assumed to play an important role.

Hence, the notion of rationality is complex and difficult to comprehend, surrounded by different voices, and situated in a web of conflict and argument as it is. The fact that the essays in this book represent the faculty of reason at work is evidence that there are limits to the concept of rationality as it is represented by *homo economicus*, a metaphor also pervading other social science disciplines.

II

In modern times, conflicting perspectives have been explored by the sociological conception of man as a role-player or a norm-guided being, in contradistinction to the economic conception of man as a calculating maximiser of utility. Nevertheless, it was the sociologist Max Weber who emphasised the distinctive role of reason in western society, particularly the role of purposive rationality. And indeed, the concept of rationality as an instrumental concept has been indispensable in economics. In other social science disciplines, purpose rationality has also come to play an increasingly important role as a basic model of human behaviour, illustrated by examples from political science and behavioural psychology.

The ongoing debate on rationality not only reflects the conflict between viewing actors as guided by norms and viewing them as calculating utility maximisers. It also gains momentum by somehow blurring the lines between viewing rationalisation as a historical process and viewing it as a defining characteristic of human conduct.

Since the ancient Greeks, the basic nature of rationality has been understood as the faculty of reasoning. And, according to Aristotle, the claim that human beings lead a thoughtful life was the particular purpose or good that distinguished humans from other living beings. The ethos of this ancient concept of rationality seems far from the currently dominant understanding of rationality as purely instrumental.

It is therefore important to make a distinction between the ethos of reason and rationality on the one hand, and the historically and culturally changing dogmas of rationality on the other. The unequivocal linking of enlightenment, modernity, and rationality sometimes falls short of reflecting the ambiguity of the historical and cultural phenomenon to which these terms refer.

After all, the notion of rationality, which, with Weber, became known as purposive or instrumental rationality, was understood from the very beginning as the result of historical and social processes, so that purposive rationality developed into the dominating form of rationality at a certain point in history. In the official interpretation, as argued by Habermas (1981), purposive rationality also became the frame of reference for other forms of action-orientation. Characteristic of purposive rationality as a type of action is that the actor or decision-maker has to consider four issues: means, ends, values, and consequences. First, the actor has to make a choice of ends, based on values, and then a choice of means, which, in turn is based on an evaluation of the consequences of using those means. The difference between "purposive rationality" and what Weber termed "value rationality" is that value-rational persons do not take into consideration the consequences of their actions. The awareness of consequences, therefore, is the *differentia specifica* of purposive rationality. As a frame of reference, purposive rationality is thus "more" rational than value rationality, because it includes the additional element of consequences. And the types of action referred to as effectual and traditional are even less rational in the context of this typology, because effectual action only includes means and ends, while traditional action only includes means.

From this interpretation of Weber's typology of action, it is easily seen that, in principle, purposive rationality does not exclude value rationality or even emotions and traditions. But it also seems clear that, according to this interpretation, value rationality is rather restricted, as it does not take into consideration the actors' responsibility for the consequences of their actions. The action-orientation included in the concept of value rationality is merely expressive of value or, as argued by Habermas, only "gesinnungsethisch." At the same time, taking Weber's typology as one's point of departure, the claim of responsibility can only be met by a utilitarian mode of reasoning. But the important point to be made here is that purposive rationality is a historically contingent achievement, and the question today should be instead: What kind of rationality comes after "Zweckrationalität?"

This question takes into account the distinction between the immanence and transcendence of reason (Putnam, 1987). The immanence of reason implies that we have no idea of rationality independent of our history, culture, language, and practice. Transcendence implies that reason also serves as a regulative idea, and is therefore a frame of reference which can be used in criticising traditions and institutions which, of course, also include specific conceptions of rationality.

Insofar as rationality is rooted in human cognitive abilities, one cannot deny rationality without denying oneself. But this does not mean that one cannot criticise certain rationalised practices and institutions that have come to dominate the rationale and style of behavioural regulation.

III

The chapters in this book all more or less reflect the complexity and ambiguity of rationality, albeit not primarily by reflecting on this ambiguity. Taken as a whole, the different discourses presented in the essays are evidence of the view that different conceptions of rationality are at play. This ambiguity is not only due to disciplinary differences, but is also to a large extent part of the disciplinary discourse itself. Since they are reflective in nature, some of the writings in this book are strongly characterised by a search for the meaning and significance of rationality as part of an endeavour to develop theories and models in the field of environmental regulation. The two chapters in the first section of the book present general perspectives on rationality, whereas the second part deals with more specific disciplinary stances on the concept of rationality.

In Chapter 1, "Rationality and Nature," Svend Andersen opens the discussion by asking what it means to act rationally when considering the impact of human actions on the environment. The chapter gives a vivid picture of the concept of "practical reason", from Aristotle to modern conceptions of rationality. The intimate connection between practical rationality and virtue is emphasised. The ability to reason in a certain way, and to make a decision in accordance with practical reasoning is phronesis. Practical reason thereby reflects the ability to derive the particular from "the normative universal" and is linked to a striving for the good.

The linking of rationality and nature is the teleological structure through which Aristotle understood living nature to also include humans, while the rational soul is what constitutes the human *differentia specifica*. Hence, there is no culture/nature divide in this approach. As argued by Svend Andersen, the rejection of the classical Aristotelian notion that cultural institutions are integrally related to human natural existence is one of the characteristic features of modern thought. Hobbes, Hume, Bentham, Smith and Kant are reviewed in developing the argument: "The cognitive end of practical reason is no longer the normatively universal, but an individual good: self-preservation" (Hobbes); "The end of practical reason is no longer cognitive, but one given by the emotions" (Hume); "Practical reason is neither cognitive nor intersubjective, but a purely subjective feeling (pleasure/pain)" (Bentham); "The individual social agent's acts are marked by self-interested means-end rationality, but the unintended aggregate result is a good for all" (Smith); and "Means-end rationality and ethical rationality are completely opposed" (Kant).

The chapter finishes by raising contemporary problems related to practical reason, and discussing Weber, Mead, Habermas and Rawls, as well as commenting on economics and rational choice theory. And the conclusion questions whether rational human action consists merely of the effective pursuit of ends and preference-based choices, or whether another form of reason exists that governs actions based on values and ethically normative concepts.

In Chapter 2, "Rationality and the Reconciliation of the DSP with the NEP," William E. Kilbourne and Suzanne C. Beckmann discuss the contemporary sociological and cultural dimensions of rationality. By viewing the environmental crisis as a crisis of paradigms, i.e., broad cultural frames of reference rather than specific values, attitudes or behaviours, the authors seek to reconcile the "new environmental paradigm" (NEP) with the prevailing paradigm of western industrial societies – "the dominant social paradigm" (DSP). Since their point of departure is the notion of rationality and the conception of the relation between man and nature as rooted in a certain cultural matrix, it becomes imperative to discuss paradigms rather than concrete methods of intervention, because the environmental crisis is believed to be a crisis of paradigms.

The chapter scrutinises the paradigmatic origin and peculiar character of the environmental crisis. The basic idea is that contemporary environmental problems are not due to failures in the past, but rather to the results of successful intentions. Arguing along the lines of Schumpeter – that it will be capitalism's success that "undermines the social institutions that protect it" – a specific discourse is developed to remedy the environmental problem. The character of this crisis is described through competing rationalities, which in principle can only be understood from within. Generally speaking, there is not one, but multiple rationalities, and to resolve these differences, the adoption of post-enlightenment and superordinate rationality is proposed.

While critically assessing the implications of the dominant social paradigm, whose core features are atomism, anthropocentrism, growth orientation and technology optimism, the authors are nevertheless convinced that some kind of merging between DSP and the new environmental paradigm (NEP), which stresses the DSP's antonyms of holism, ecocentrism, steady state and low technology, will have to take place if the environmental crisis is to be remedied.

Chapter 3, "Rationality, Institutions and Environmental Governance", by Martin Enevoldsen, focuses on the means of environmental governance. It addresses the rationality question from the intersection of political science and economics, investigating the theoretical basis for designing effective instruments for environmental governance.

Concepts of rationality have played a central role in the development of modern environmental governance, especially qua rational choice theory. The first perspective on environmental governance, formulated on the basis of the assumptions of rational choice theory and the associated methodological individualism, is the Pigouvian externality theory. It focuses on the negative side effects of polluting enterprises in a diffuse open market. A second perspective is the collective action theory, focusing on strategic dilemmas associated with the "tragedy of the commons". Both theories draw attention to the contradictions between individual and collective rationality, and the need to address the problem by way of pollution-control instruments, but the kinds of problems on which they focus, and the instruments they recommend, are quite different. They are contrasted with political institutionalist theories on environmental governance – policy implementation theory and the Berlin school – which have offered solutions to problems that could not be handled by the instrumental rationality approach.

A history of environmental governance theory has not yet been written. Part of the reason is the overlapping of economics and political science in this field of research. Another reason is that the theory and practice of environmental governance is still very young. While chapter 3 of this book does not pretend to make up for this gap in the literature, it nevertheless presents a comparative theoretical survey, including both the instrumental rational approach, which has mainly been advanced by environmental economists, and the institutionalist approach developed by political scientists interested in environmental politics. Hence, in this chapter the meaning, importance, implications and limits of instrumental rationality are discussed through a critical survey of the principal theories on which modern environmental governance has been built.

In Chapter 4, "The Ethical Rationale of the Concept of Sustainability and Rationality Conflicts in Environmental Law", written by Ulli Zeitler and Ellen Margrethe Basse, the concept of sustainability and the implementation of sustainability in environmental law are the central issues. Environmental problems are regarded as a global phenomenon, but since the world consists of different cultures, the implementation of regulatory devices is bound to reflect this condition. The chapter thus examines differences between legal and ethical rationalities, differences which also substantiate divergent interpretations of sustainability in legal and political documents, and which are responsible for frictions and dilemmas in environmental management. The discussion begins by comparing legal notions and rationalities in Europe, New Zealand (integrating European and Maori culture), and Japan, with major differences being observed among these cultures. In an attempt to cope with these differences, the authors introduce a distinction between first-order rationalities, concerned with basic ways for human beings to structure experiences and form life expectations, and second-order rationalities, as expressed in different moral theories. The first-order distinction entails two major discourses: an anthropocentric versus an ecocentric way of thinking, and reflective rationality versus non-reflective (bodily) rationality.

The application of the above terminology brings to light important differences between the three chosen cultures when it comes to understanding the meaning of sustainability and the implementation of environmental law. This also emphasises that different historical and cultural conditions constitute obstacles to any unanimous interpretation and implementation of sustainability.

The chapter concludes with a discussion based on the Japanese experience, according to which the distinction between anthropocentrism and ecocentrism gives no meaning at all. The enforcement of a global consensus on the perception of "nature," "environmental problems" and the "quality of life" is seen as both morally reprehensible, unrealistic, and inefficient. Emphasising local thinking and action, the authors criticise the ecological movement for having propagated the motto "think globally and act locally," since there are obvious limits to a credo of global thinking.

In Chapter 5, "Rationality Deficits in Behavioural Intervention Strategies," Erik Kloppenborg Madsen and Folke Ölander take their point of departure in the fact that different intervention strategies reflect differences in assumptions concerning human reasoning and motivation. Differences in modes of reasoning and motivation reflect the impact of societal and cultural factors on actions and decision-making, which, in turn, may give rise to a discussion of different assumptions about rationality. But if such differences are taken for granted, the rationality of a regulatory device is also at stake in at least two respects. Firstly, an intervention may simply be ineffective if based on the wrong assumptions about the individual's motivational structure, and secondly, the means used may have unintended side effects, outweighing the short-term benefits of an immediate change of behaviour. Among the unintended side effects mentioned are diminished personal responsibility and ability to cope, as well as a weakening of civic society and democratic procedures.

The discussion is centred on two often-competing models of reason: instrumental rationality and norm-guided rationality. In line with the distinction made by March and Olsen (1995) between two perspectives on governance, "the exchange perspective" and "the institutional perspective," respectively, it is further argued that the institutional notion of "appropriateness" may be developed in such a way that it takes into account not only undeliberated values, but also principles and norms based on reasoning and arguing.

There is a need to discuss regulation, reflecting on the fact that the instruments of change not only alter behaviour, but also people and the relations among people. A fundamental challenge of regulation in a pluralistic society is the reconciliation of collective and individual values, beliefs and interests. In order to meet this challenge, the authors propose that the institution of reasoning and arguing is to be seen not only as a means of persuasion, but as one of the primary aims of intervention.

The aim of Chapter 6, "Rationality, Environmental Law and Biodiversity," is to consider and discuss how decisions are influenced by a number of different factors in the legal system. Helle Tegner Anker and Ellen Margrethe Basse examine a string of issues linked to the problem of environmental regulation: "Rationality in law and jurisprudence," "Environmental law and the concepts of rights and interests" and "Biodiversity, law and rationality."

The authors demonstrate that two different categories of rationality are at stake, particularly in relation to environmental law. They argue that the first category reflects traditional norm-oriented as well as goal-oriented rationality, while the second category of rationality relates to deliberation, fairness, equity and the virtues of clarity and open-mindedness, and thus to notions of reflexive law and communicative rationality. Due to the dynamics and changes in modern society, they call for the rethinking of jurisprudential concepts such as the sovereignty of states, property rights and responsibility, because the protection of common resources cannot be guaranteed by traditional interpretations of these notions.

The condition of the environment challenges ethics and jurisprudence by giving rise to the questioning of current perceptions of the rights and interests of future generations, endangered species, civic duties and the like. The concept of interest in environmental law – it is argued – may have a meaning at variance with the traditional legal concept of interest. According to the authors, the protection of such interests may require a brand of rationality based on public deliberation rather than a cost-benefit analysis. The interests in question are vague and imprecise, and demand a rationality of conduct that allows for the virtues of reflection, consideration and responsibility.

The chapter concludes with an example of the complexities of environmental law represented by the Convention on Biological Diversity. The authors demonstrate that the complexity of environmental law most likely calls for a species of mixed rationality, but at the same time they emphasise the need to supplement traditional norm- and goal-oriented rationality with a rationality based on deliberation and communication.

The discussion and conclusion of Chapter 7, "When Choice of Means Undermines the Goals: Rationality from a Psychological Perspective," is based on results from empirical psychological research into social dilemmas. In discussing the concept of rationality, John Thøgersen and Tommy Gärling use rational choice theory as their frame of reference. Demonstrating that decision-making frequently moves away from rationality in a strict rational choice sense, they argue that rational choice theory describes a logic of choice rather than a psychology of value. However, this does not (nor do they try to) explain how and why different values are prioritised. Therefore a theory of values is called for.

After having discussed revisions and extensions of rational choice theory pertaining to the accommodation of the model to empirical findings, the authors discuss the topic of individual versus collective rationality. Collective rationality is defined as the extent to which groups effectively pursue common goals. This is the field with which social dilemma researchers have been occupied for some time. In discussing factors promoting rational collective decision-making, the authors take up the notions of social and personal norms as important factors in enhancing collective rationality. The conflict between individual consumption and environmental protection is addressed as being an instance of a conflict between individual and collective goals. The authors discuss several means of promoting restraint on individual self-interested rationality. Generally, all means of regulation have strengths as well as weaknesses.

Special attention is paid to the discussion of intended effects versus the negative and positive side effects of using coercive and non-coercive means, respectively.

Coercive means may have positive side effects because authorities convey a strong signal of commitment to solving the targeted problem. This, however, may also be perceived as an infringement on individual freedom, which could ultimately result in psychological reactance, and thereby create a negative side effect. Positive side effects of non-coercive means are likely to occur if people attribute their restraint to internalised attitudes and norms. Negative side effects of non-coercive means may occur if the message conveyed is perceived as having low priority.

Although the most radical way of solving most environmental problems would be to abolish the social dilemma itself (i.e., changing the structural characteristics of the decision-making situation in such a way that individual and collective rational behaviour converge), the authors conclude that such a strategy is impossible. This is the case, not only because of a lack of knowledge and control, but also because such a strategy would be inconsistent with values of democracy and equity.

IV

The multidisciplinary views on environmental regulation and rationality presented here all concur in the importance of understanding that the notion of rationality is complex and characterised by a wealth of different and diverging connotations. The essays collected here are meant to provide the basis for further efforts to deal with rationality and regulation in the environmental arena, both from a normative and an empirical perspective. The normative perspective is based on the assumption of rationality (or rather: rationalities?) providing a model (or rather: models?) for human identity and action, from which appropriate procedures for obtaining human objectives can be derived. The empirical perspective manifestly offers interpretive, explanatory and predictive accounts of human action.

Ultimately, any understanding of rationality requires an approach integrating individual-level analysis with the analysis of macro-phenomena. Such an approach takes into account that rational actors both create and are constrained by the societal rules embedded in norms and institutions, and, moreover, allows us to assess the diversity of rationalities at play, including their conflicts and compatibilities.

REFERENCES

Habermas, J. (1981). *Theorie des kommunikativen Handelns.* Frankfurt a.M.: Suhrkamp.

Putnam, H. (1987). Why reason can't be naturalized. In: K. Baynes, J.Bohman, and T. McCarthy (Eds.), *After philosophy – End or Transformation?*, pp. 222-244. Cambridge, MA: MIT Press.

Rawls J. (1993). *Political Liberalism.* New York: Columbia University Press.

Sagoff, M. (1988). *The economy of the earth.* Cambridge: Cambridge University Press.

Schumpeter, J. A. (1942). *Capitalism, socialism and democracy.* New York: Harper and Row.

Part I

CHAPTER I

Rationality and Nature

Svend Andersen

WHAT IS RATIONAL action insofar as our actions impact on the environment? This is surely one of the most fundamental questions facing the scientific study of the environment – a question ranging under "the philosophy of social science", and one that conjoins environmental research with environmental ethics. The concept of "rational agency" is extremely complex and contentious. One way of acquiring a grasp of its various possible meanings is to review its historical development. In the following I will aim at outlining some elements of European thought on practical rationality. Since our interest is in examining the nature of human agency in relation to the environment, the various conceptions of practical rationality will be linked to matching conceptions of nature and human attitudes towards our natural surroundings.

THE CLASSICAL CONCEPTION OF PRACTICAL REASON

The most obvious place to start is with Aristotle's ethics. According to Aristotle, ethics is a matter for reason, a concept made evident in his definition of the concept of "happiness" or *eudaimonia* as the human good. The definition is that the good is the activity of the soul (EN[1], 1098a, 17). Immediately prior to this, Aristotle asserted that the human function is the soul's activity "in thinking" or "according to rational principles" (*kata logon*). In the definition of eudaimonia quoted, he adds "in accordance with virtue". This brings us to one of the most critical features of Aristotle's ethics and one of the most difficult to construe

1. EN = Ethica Nichomachea.

correctly: the intimate connection between practical rationality and virtue. This relationship is also manifest in the celebrated distinction between the moral and intellectual virtues, which again builds on the distinction between the rational and the irrational "parts" of the soul. But reason is also important for those moral virtues that link to the irrational soul, for the morally relevant aspect of the irrational is what is accessible to reason: feelings, passions, affects – as opposed to the vegetative functions, which do not admit of formation and translation into conduct. The moral virtues too, are rational, which is why Aristotle's analysis of them reaches its conclusion only when the action-related intellectual virtue of phronesis (judgement) has been examined.

But how is it possible, according to Aristotle, for reason to influence action, and what is the relationship prevailing between theoretical and practical reason? The answer can be sought in the introduction to Book VI, where the inquiry into intellectual virtues begins. Aristotle distinguishes between two aspects of the rational soul. One aspect of reason is concerned with scientific cognition (episteme); its concern is with the rationale or causes of things, and the specificity of scientific knowledge is that the causes in question are immutable. The other aspect of reason is directed towards the changeable, which is to say those cases where the causes of change are to be sought in the human person. They can be thus sought in two ways: either in a ground that creates (poiesis) or in a ground for action (praxis). The difference between these two basic forms of human activity is that the aim of poiesis is to produce an artefact, whereas that of praxis is that action does not exist for the sake of something else – it is in itself a part of the good life.

In Aristotle we find two types of action-related reason complementary to theoretical cognitive reason. One I will call technical reason, since Aristotle terms the disposition towards the proper unfolding of poiesis "tekne". The other we may call practical reason, since it is associated with praxis, moral agency. In order to more precisely define the nature of practical reason, let me first briefly review what Aristotle understands by theoretical reason.

Theoretical reason, as already noted, is expressed in scientific cognition. Its specificity is its ground, and its ground derives from the relation between the universal and the particular. Reason is not the antithesis of the senses, and cannot function independently of them. To Aristotle, the senses are the instrument for registration of the individual or the particular, but are also potentially a cognition of the universal, represented by the individual. Reason contributes the actual cognition of the universal. The rationality of scientific cognition finds expression in demonstration, which for Aristotle is famously syllogistic in form. Theoretical reason then manifests itself in its grounding of knowledge through the derivation of specific or distinct propositions from universal ones. But Aristotle makes the important assumption that the progression from universal propositions to those at an even higher level of generality is not infinite. Highest-order universal propositions, i.e. principles, do exist. Theoretical reason presumes a determinately ordered system of universal truths.[2]

Practical reason in Aristotle displays the same structure as does theoretical: it consists in the ability to derive the specific from the universal. But now the specific is not simply the perceived individual, but the concrete act. Time and time again, Aristotle stresses the fact that praxis is specific or distinct. But what then is the universal? The universal to which practical reason is related is obviously not purely factual. In the first instance I shall refer to it as the normatively universal. The operational intellectual virtue in the context of determining what act is to be performed in the concrete situation is, as already indicated, phronesis, or judgment. It finds its expression in the practical syllogism. However, interestingly enough, when Aristotle gives examples of the practical syllogism in the Nichomachean Ethics, they never concern moral action as such. He draws his examples from the art of medicine, for they concern the benefits of eating pale meat and dry food, and avoiding drinking heavy water, with the fullest account touching on the second example. The practical syllogism here seems to take the following form:

Major premise: All kinds of dry food are good for all humans.
Minor premise 1: I am a human
Minor premise 2: This piece of bread is dry food.
Conclusion: I should eat this piece of bread.[3]

The ability to reason in this manner is, as already noted, phronesis. As a rule, Aristotle stresses that phronesis is the capacity to identify the correct means of achieving one's chosen goal. This is to be understood as meaning that phronesis is primarily the ability to apply the general knowledge expressed in the major premise in a specific situation. One

2. For the foregoing, viz Höffe 1976.
3. Here I follow Rackham's translation and commentary in the cited edition.

might therefore be led to conceive of phronesis as the simple ability to find the correct means of achieving a stated goal, or the purely formal ability to apply a normative universal proposition to a concrete situation. We might call this type of reason neutral means-end rationality. And Aristotle does indeed acknowledge this kind of reason, but calls it, not phronesis, but *deinotes*, cleverness. Cleverness is precisely the neutral ability to find the right means of attaining the goal one happens to have set oneself. Cleverness in itself is neither noble nor base; whether it is the one or the other depends on the goal. Phronesis includes cleverness, but both the cunning person and the possessor of phronesis may be called clever. Indeed, even a vice like akrasia (intemperateness) is compatible with cleverness.

An understanding of phronesis that even more readily presents itself is that it is self-interested means-end rationality. As is his wont, Aristotle proceeds in the manner of "linguistic analysis" and starts from what a given word, *in casu* phronesis, means in ordinary language. And it seems primarily to have denoted something in the order of the individual's ability to realize his/her own good. But it is not this meaning of the word that defines phronesis as a virtue, i.e., as moral reason. Phronesis, in its original sense, is the capacity to identify that act which, in the specific situation, realizes the true good and the good of all: in essence, the ability to draw a practical inference from the normatively universal.

But in what sense is practical reason normatively universal, grasping the true good of *all* people? This is where the close connection between the moral virtues and phronesis is made manifest. Rational wisdom concerning the normatively universal is not knowledge of a merely formal universal rule of action. The concept of rational agency in that sense is a modern one, to which I will return. To Aristotle, practical reason is linked to a striving for the good. But this striving is no spontaneous, natural phenomenon – no kind of sovereign manifestation of life. The striving for the good is "right inclination", and it qualifies as right only if linked to a principle of reason (logos) (1139a, 30). In other words: a person can only have (universal) knowledge of what the good life consists in if he/she is good, i.e. in possession of moral virtue (1144a, 7ff.). And knowledge of the true good is conditional on the ability to practise it. Knowledge of the normative universal is at once both the cultivated ability to achieve the mean which is common to all virtues, and rational knowledge of the right principle (1144b, 26ff.).

Let me briefly indicate features that are conducive to an anachronistic reading into Aristotle's text of present-day problematics.

Firstly, it is tempting to conceive virtue ethics as a communitarian version of ethical relativism. Virtues are acquired by the individual through practical experience and induction into the membership of a given community. But concurrent with the development of particular dispositions to act in specific ways (individual moral virtues), the individual gains knowledge of what the good is to human beings. As ethicist, Aristotle is both cognitivist and realist.

Secondly, Aristotle makes the assumption that economic and political life is governed by practical reason. Aristotle makes an express point of saying that both political science, i.e. statecraft, and the running of a household are forms of phronesis. Particularly as concerns commerce, Aristotle holds that exchange and other financial transactions should conform to a structure corresponding to that of justice (EN V, 1132b, 34ff.).

Thirdly, it is tempting to translate certain of Aristotle's statements on phronesis into the currency of contemporary thought on self-interested means-ends rationality or prudence (*prudentia*). The reality is probably that Aristotle is innocent of any sharp distinction between the individual's interests and the common good, or between egoism and altruism. This appears in his analysis of friendship, according to which righteous self-love is *eo ipso* love of others. The single-minded pursuit of self-interest constitutes a moral defect – a lapse from practical reason (EN IX, 1168b, 14ff).

In Aristotle's philosophy, practical reason is linked to human attitudes to nature in various ways. We have already touched on the fact that (1) practical and theoretical reason share the same structure, i.e., that structural congruence obtains between human acquisition of knowledge of nature and human agency in the social sphere. But this agreement goes deeper, since the individual as a social agent is part of nature, and thus not without some sort of kinship to non-human nature. This primarily manifests itself in two ways.

(2) Even though reason, according to Aristotle, sets humans apart from other (sublunary) natural beings, the possession of reason does not prevent the human person from being precisely a natural creature, a zoon. The human individual's social or political nature, too, is a property of which he/she is possessed in virtue of being a natural being. Reason is the specific form of psyche ("soul") which distinguishes humans from other living things. But precisely because the human being is a rational animal, he/she shares aspects of psyche, i.e. the life principle, with other living things. At the lowest level, humans exhibit the

same life-functions as do plants and animals – namely in terms of what we would call metabolism; these functions Aristotle attributes to the vegetative soul. At a higher level there are life-functions manifested only by humans and animals, namely those belonging to the sensitive soul (sensation, perception and primal urges). Only the rational soul is distinctively human, constituting the human *differentia specifica.*

Ultimately, the concord that obtains between human practical reason and nature is grounded in (3) Aristotle's teleological conception of nature. Both poiesis and praxis are governed by nature, in the sense that reason is the ability to identify the right goal (*telos*), and to select the appropriate means for its realization. In that sense practical reason is teleological. But in virtue of so being, human reason is subsumed under a comprehensive and cosmic pattern. For it is a fundamental feature of nature and cosmos, according to Aristotle, that each and every entity possesses a characteristic property determining its activity. This property too, Aristotle terms "nature" (*physis*), so that in Aristotle we already encounter the duality of semantic content that the concept displays today: "nature" can mean both the sum of natural phenomena (nature) and a thing's essential properties (e.g. human beings' nature). A thing's essential property determines what it strives towards, its telos. Thus a stone seeks the natural home of the element earth, the earth's surface, whereas a human being seeks to realize his/her potential through living a good life. In Aristotle, the laws of motion in physics, and the theory of agency in ethics display the same structure. As agents, human persons are part of nature, and the institutions in which their lives are played out – the household (*oikia*) and the state (*polis*) – are founded in natural communities. A conflict between human rational action and nature is inconceivable within the contours of Aristotle's world picture.

MODERN CONCEPTIONS OF PRACTICAL REASON

The rejection of the Aristotelian notion that human social institutions are integrally related to human natural existence is one of the characteristic features of modern thought. This rejection can clearly be seen in Thomas Hobbes: according to him, it is by no means a naturally given thing that we join with others – certainly not in the sense that we do so on the basis of a feeling of "good will" towards them. The foundation of society is not reciprocated benevolence, but "love of ourselves" (DC

1,2[4]). For Hobbes it is not a case of continuity between human social life and the so-called state of nature, but rather of opposition. This opposition marks both the new science and the corresponding philosophical understandings of the relationship between human beings and nature. Both aspects are exemplified in the work of Rene Descartes. He was, like Galilei, an exponent of the new science (primarily physics and astronomy) whose distinguishing features are empirical experimentation and mathematical description. The rationality that typifies acquisition of scientific knowledge of the physical world can thus be designated mathematical-computational. However, this fact gives rise to an opposition between nature as object for the investigative human subject, and this very same subject's own nature. Human beings are, to Descartes, essentially consciousness and thinking (*res cogitans*), whereas observed nature is extension and thus materiality (*res extensa*). The human body belongs to this last category, so that the conception of human beings answering to the modern view of nature is dualistic. The human person, according to Descartes, is not his/her body; nor is he/she, strictly speaking, a part of nature. This scientific and philosophical view neatly dovetails with the idea that human social life stands in contrast to human natural existence.

Hobbes

Turning once again to Hobbes, one might surmise that he would pursue a similarly radical revision of practical reason by, for instance, conceiving it in terms of self-interested means-ends rationality. But this is not actually the case. Admittedly, Hobbes has adopted the idea of the mathematical character of reason, so important to the natural science of modernity. Thus he seeks to derive logical functions from the arithmetical functions of addition and subtraction (Hobbes 1983, 81f.). But when it comes to the structure of practical reason, his thinking is strikingly traditional. One commentator goes so far as to say:

> Hobbes' concept of reason has more in common with the classical philosophical tradition stemming from Plato and Aristotle than with the modern tradition stemming from Hume. (Gert 1991, 13).

---

4. References to Hobbes 1991 are shown by the abbreviation DC, followed by numbered chapter and paragraph.

In point of fact, practical reason is, for Hobbes as for Aristotle, the ability to apprehend the human good; it is cognitive in relation to the normative. The novel in Hobbes is his conception of what the human good really is. In other words: the novel is that reason basically stands in relation to evil. It has been stressed that what makes Hobbes the first modern political philosopher is the role he assigns self-preservation.[5] Self-preservation becomes a socially constitutive factor because each individual knows that any and every other person can bring about one's end. Every individual's efforts are directed towards the preservation of life and limb. Such efforts are rational. Their aim is something future ("things to come") and thus cannot be an object for the senses, whose deliverances belong to the present (DC III, 31.). The laws of nature, which, according to Hobbes, are formulated in conjunction with the social contract, are founded in "right reason". Right reason comprises both a sound apprehension of what humans seek (self-preservation and through it, peace) and the choice of appropriate means towards securing this end.

Practical reason is still, in Hobbes, the capacity to cognise the normatively universal. The modern in his view is the idea that normativity is constituted through the future possibility of individual annihilation. This possibility applies to one and all. In consequence, it becomes the foundation for social normativity, in that self-preservation is best secured through peace as a common project. The individual accepts common rules as binding by renouncing the exercise of his/her natural rights – to the extent that others do likewise. This is Hobbes' interpretation of the Golden Rule (Hobbes 1983, 147). Since it is by virtue of right reason that the individual engages in the social project, practical reason in Hobbes takes the form of the Golden Rule.

Hume

In Gert's view a modern conception of practical reason is to be found, as mentioned above, in David Hume. Hume formulates it clearly and crisply as follows:

> Reason is, and ought only to be the slave of the passions, and can never pretend to any other office than to serve and obey them. (Hume 1978, 415).

5. Viz Henrich 1976.

This can also be expressed by saying that reason, in Hume's opinion, has merely a theoretical function. Theoretical reason may well be practically relevant, but reason cannot be practical in the sense of motivating or initiating acts. The motivational force behind our acts can only be the passions. They admit of the influence of reason in two ways. Reason may determine whether the occurrence of a given passion is warranted or not. If, for instance, one fears a certain thing and reason ascertains that it does not exist, it removes the basis for the fear. Further, reason, with its knowledge of causal connections, can identify the appropriate means of enabling a passion to achieve its goal. (416).

No more can moral motivation derive from sources other than the passions. Our passions include, according to Hume, "calm passions" such as "the general appetite to good, and aversion to evil, consider'd merely as such" (417). It is this calm passion he also designates as moral sentiment. In contrast to the feelings adhering to particular relations to others, e.g. aversions, which are ultimately expressed in "the language of self-love", a person's moral judgments are expressions of feeling, expressive of "a point of view, common to him with others":

> ... the humanity of one man is the humanity of everyone; and the same object touches this passion in all human creatures. (Hume 1987, 75).

There is then, in Hume, something normatively universal. But our experience of it is not cognitive; it is not a matter for reason, but for feeling. Insofar as one may speak of practical reason in Hume, it must be in terms of a neutral means-end rationality. Rational reflection may indeed end in a conclusion that points to a particular act. But, as MacIntyre notes, the conclusion does not in itself lead to action. The reasoning is marked by a hypothetical condition: if the act is willed, it should be done (viz. MacIntyre 1988, 303f.).

Bentham

In its original version in Bentham's writings, utilitarianism retains Hume's basic conception of reason's practical role. However, here it is given a characteristic twist, as reflected in the following formulation:

> Passion calculates, more or less, in every man. (Bentham 1996, 174)

Reason is a tool for the passions which impel human action, but reason's practical function is now primarily calculation. This must be understood in the light of the claim that the passions can ultimately be traced back to two feelings: "pain and pleasure" (11). Since these feelings have quantitative dimensions (e.g. intensity and duration), it is possible to calculate the net sum of pleasure. As we recall, this feature serves as a fundamental assumption for the application of the principle of utility, which in Bentham's version has it that acts must be judged on the basis of their tendency to increase or diminish the happiness, i.e. the sum of pleasure, of the parties concerned (12). The principle of utility may be regarded as emblematic of that state of affairs where a conception of the human good valid for all no longer prevails. What is common to all is no longer an end in the sense of a telos which dictates what the good life is, but rather a "metron" which makes it possible to determine the amount of pleasure in terms of which each stretch of experience can be expressed. Ethical reason has now become an economically defined means-end rationality.

Bentham also regards the principle of utility as rational, in the sense that it can be argued for by showing that it is superior to all other principles, not least by being internally consistent (15ff.). But what becomes of morality's universality? A standard objection to utilitarianism is that, precisely in virtue of its calculative aspect, it leaves open the possibility that an individual may be sacrificed to secure the greater advantage of others. However, it cannot be denied that the principle of utility has its own form of impartiality. It comes to expression in what has been called Bentham's dictum, which John Stuart Mill cites in the context of his account of justice: "everybody to count for one, nobody for more than one" (Mill 1985, 319). In Mill's view, the impartiality articulated here inheres in the principle of utility itself: one person's happiness is to count just as much as the next person's – given that the quanta of happiness in each case are equal. Impartiality lies in the fact that the happiness of each of those parties affected by a particular act perforce figures in the pleasure calculus. This does not prevent the individual from losing out, for his/her loss has to be set off against the gains of the others.

Adam Smith

The conception of practical reason that might be said to be contained in Adam Smith's economic theory is in important ways akin to

Hobbes's. Where Hobbes stresses the individual's needing to turn to others merely as regards the preservation of life, Smith emphasizes in a more general way mutual dependence as a fundamental condition of life: the individual is dependent on help from others. And, says Smith, the most effective way of ensuring others' help is not through an appeal to their benevolence, but to their self-love. In commerce we obtain goods from others by drawing on their self-love:

> It is not from the benevolence of the butcher, the brewer, or the baker, that we expect our dinner, but from their regard to their own interest. (Smith 1994, 15).

A noteworthy feature of Smith's conception of social action is that the sum of individual self-interested acts comes to constitute an unintended order. This is the idea he expresses with the well-known image of the invisible hand. He makes use of it to describe the circumstance that despite an investor only being interested in acquiring maximal returns, he contributes through his acts to the maximization of the nation's assets:

> By pursuing his own interest he frequently promotes that of the society more effectually than when he really intends to promote it. (485)

The radical import of this insight lies in the fact that the aggregate result of the acts of individuals is not intended by any agent, and yet appears precisely as though it were. By acting from a self-interested, maximizing means-end rationality, the individual brings, so to speak, social structures into being which bear a systematic aspect. Adam Smith is thus one of the first to use individual means-end rationality as a model for the explanation of social processes. Before pursuing this line of thought further, it is necessary to mention Kant's contribution to modern thinking on practical reason.

Kant

The modern alternatives to Aristotle's conception of practical reason I have dealt with until now are really merely modifications of it. They are various: practical reason may be assigned a cognitive content (Hobbes); or reason may be denied normative cognitivity and viewed simply as a tool (Hume, Bentham, and Smith). We find in Kant a radical departure from this mode of thinking. He introduces a wholly new conception of practical reason, namely that which manifests itself in the *categorical imperative*. The radical nature of his break with Aristotle comes to expression in the fact that he regards both poiesis-rationality (*techne*) and praxis-rationality (phronesis) as hypothetical imperatives and dismisses them as irrelevant. Peculiar to the categorical imperative is that the universally normative is its object, insofar as it commands it. The unconditional moral "you must" has universality as its content. This appears to be what the first formulation of the imperative says:

> Act only on that maxim through which you can at the same time will that it should become a universal law" (Kant 1785, 52) (Paton's translation).

Kant would thus seem to be the first to formulate the so-called principle of universalizability: "If an act is right for one person in a given situation, the same act is right for any other person in similar circumstances". But there can be no doubt at all that the categorical imperative amounts to far more than a rule of this type. Its fundamental point is better expressed in a formulation like the following: "You must represent humanity in your acts". Only on the basis of such an interpretation does it become clear why Kant asserts that the same imperative may be expressed alternatively:

> Act in such a way that you always treat humanity, whether in your own person or in the person of any other, never simply as a means, but always at the same time as an end (66f.) (Paton's translation).

In obeying this imperative the human person manifests him/herself as a free being, capable of determining his/her acts independently of any given natural impulse. That reason is practical in this sense cannot be grounded further; it presents itself as a fact (Kant 1788, 72).[6]

Modern conceptions of rationality in summary

The various configurations of practical rationality outlined in the foregoing can be summarized in the following theses:

---

6. Ernst Tugendhat thinks Kant's concept of practical reason conflicts with what we normally understand by rationality, so he designates it: "Vernunft-fettegedruckt" (Tugendhat 1993, 45).

1. The cognitive end of practical reason is no longer the normatively universal, but an individual good: self-preservation (Hobbes).
2. The end of practical reason is no longer cognitive, but one given by the emotions; but although the end is now made subjective, it is still intersubjective (Hume).
3. The end of practical reason is neither cognitive nor intersubjective, but a purely subjective feeling (pleasure/pain); however, calculability and maximizability are trans-subjective (Bentham).
4. The individual social agent's acts are marked by self-interested means-end rationality, but the unintended aggregate result is a good for all (Smith).
5. Means-end rationality (the hypothetical imperative) and ethical rationality (the categorical imperative) are completely opposed (Kant).

CONTEMPORARY PROBLEMS CONCERNING PRACTICAL REASON

Kant's clear separation of means-end rationality and ethical rationality presupposes Cartesian dualism, i.e. the assumption that the human person as an ethically rational being is not part of nature. When agents are driven by feelings and other natural impulses, they are quite capable of obeying hypothetical imperatives, and thereby of acting in accordance with means-end rationality, but such agency is not ethical. For ethical agency, for Kant, is acting for the sake of duty in the sense of obedience to a norm-governing agency, and agency of this kind presupposes freedom in relation to those causal contexts in which empirical impulses figure. Any such freedom is only possible if the human person as a rational being owns another form of existence than that given by nature, namely what Kant calls "intelligible" existence. The dignity (Würde) the human person enjoys as an ethical subject means that he/she is separate from and above nature. Kant's rationalist ethics is thus both anthropological and idealistic: it presupposes that it is possible to describe and conceive of the human person in categories that are quite different from those of science.

This idealistic presupposition loses its persuasive power in the light of Charles Darwin's theory of biological evolution. This theory may be seen as eliminating what remained of the Aristotelian-teleological conception of nature. Its consequence is that the apparent purposiveness characteristic of organic nature can be traced back to non-teleological regularities. Living organisms did not arise, according to Darwin, because they were "designed" to pursue specific ends. On the contrary, they have evolved on the basis of three fundamental principles: genetic change resulting from mutation, variation in respect of adaptation, survival and selection of the best-adapted individuals.

The conception of nature to which the theory of evolution leads has two important consequences for the conception of the human person as a rational agent. Firstly, the idealistic assumption that humans belong to another order than merely that of nature is under attack: the theory implies "the descent of man". Secondly, the conception of organic nature is non-teleological: species do not emerge to fulfil a particular function, and thus the principle behind evolution is contingency. On the other hand, there is, apparently, a kind of rationality inherent in the natural world. The adaptation of organisms to their environment (in the form of the dissemination of the genetic pool) can be described as though these organisms were exhibiting a means-end rationality.

Contemporary conceptions of nature are marked by the generalization of the evolutionary perspective, in that the entire cosmos is thought to have undergone a process of development, starting with the "Big Bang" and moving towards an ever-increasing state of disorder (entropy). On the philosophical level, twentieth century science has, however, moved towards a recognition of the limitations of rational explanation. This has meant abandoning the idea (scientism, reductionism) that all aspects of reality are susceptible of scientific description. Specifically as concerns organic nature, there appears to be a certain tension in science between genetic-evolutionary biological approaches and an ecological approach. Ecology opens up for a holistic understanding, in the sense that it stresses the coexistence of manifold organisms in a supra-individual biological whole (ecosystem, biotope.)

Instrumental and value-rationality (Weber)

The development in empirical research in the nineteenth century exemplified by Darwin led to a bifurcation of the concept of science, which finds expression in German in the distinction between "Naturwissenschaft" and "Geisteswissenschaft". Paralleling this distinction between areas of research is the methodological distinction between "Erklären" (explaining) and "Verstehen" (understanding). In this same period new scientific disciplines emerged, not least as a result of subject areas being separated from philosophy. Psychology and sociology are cases in point. The question was whether these subjects, addressing, as they do, the issue of human agency, belong to the natural or the social sciences. A relevant classical theory of practical reason here is that of Max Weber.

Weber's theory of practical reason is rooted in a more comprehensive set of problems – that concerned with the relation between modernity and rationality. The outline of Weber's theory given in the following draws on Jürgen Habermas's account. There are two distinct elements in Weber's theory: (1) a theory of rationalization in the sense in which that term denotes a particular development in European (occidental) civilization, (2) a theory of rational agency.

(1) The course of development that Weber calls modernization and rationalization is integrally linked to a process of differentiation, i.e. the respective isolation of various spheres of life which once constituted a whole. In this context the distinction made by Parsons between three basic spheres: society, culture, and personality can be applied (cf. Habermas, 1981). Further differentiation takes place within each sphere. Thus it comes about that the capitalist economy of a society is separated from the centralized power of the state, and both these spheres or subsystems are closely linked to the institution of law. In culture, science and art figure as separate phenomena: they are rationalized in the sense that science becomes mathematical and experimental art is characterized by its authenticity. Habermas sums up rationalization in a definition: rationalization for Weber is

> ... jede Erweiterung des empirischen Wissens, der Prognosefähigkeit, der instrumentellen und organisatorischen Beherrschung empirischer Vorgänge. (Habermas 1981, 228).

Rationalization is thus characterized in terms of planning, predictability, repeatability and control. Governmental bureaucracy and accountancy in the financial world typify rationalization.

(2) Society and culture are collectivities, but they are made up of individuals and their actions. There is, then, a connection between rationalization in the various sectors of society and individuals' rational agency and ways of life. With regard to the latter, Weber introduced what is now a celebrated distinction, namely that between (i) Zweckrationalität and (ii) Wertrationalität. Re (i): means-ends rationality denotes, as the expression itself indicates, the capacity to orientate one's action in relation to ends, means and side-effects, so that means are balanced against ends, ends against side-effects and diverse ends are balanced against each other. A means-ends orientated act of this kind is quite dis-

tinct from spontaneous or traditional practices. Means-ends action requires the mastery of a technique, i.e. the ability to apply ends in a regulated way. Method-governed agency may be encountered in all kinds of activities, even in the religious sphere; for instance, there are such things as techniques of prayer! Means-ends action may take the form of intervention in the environment, and the criterion of rationality is related to its effectiveness. Re (ii): when there is talk of rational agency today (e.g. in the social sciences), it will often merely be the individual's interests or preferences that are appealed to as determinants of the ends of action. Weber holds, however, that a person's preferences are grounded in values. By value-rationality is meant the assertion that choice of ends is founded on values. Value-rational agency is what is brought into play when, without regard to consequences, the individual acts from a belief as to what duty, dignity, aesthetic standards or one's religion dictates. To say that values are rational is to say that they make possible a way of life governed by principles.

The two types of rationality – end- and value-governed – jointly make the leading of a methodically ordered life possible. Weber calls such a form of life an ideal type; the earliest approach to it, he believes, is to be found in what he terms the "calling-asceticism" of Calvinist and puritan sects. Here then, lies the heart of his celebrated thesis on the connection between the spirit of capitalism and Protestant ethics: in certain forms of Protestantism a connection may be traced between a methodical and rational way of life, e.g. between a commercially enterprising life and the religious belief in being predestined by God to salvation.

Intersubjective rationality (Mead)

As we have seen, an important feature in modernity's conception of practical reason is that the subject of practical reason is the individual. In consequence, the existence of a shared rationality requires a specific explanation, e.g. the idea of a social contract or the coordination of individual agency via an "invisible hand". It is true to say, then, that in modern thinking the idea that rational agency might be combined with the application of a common or impartial point of view has been problematized. With Aristotle it was otherwise; he regards it as unproblematic to speak of the common good.

To adopt the impartial point of view is to regard oneself as being like all others. Yet it is precisely a constituent of modern experience that I am in fact, different from all others. The dilemma dissolves if it can be shown that that there is no contradiction between being an individual, a self, and being like all others. Hegel sought to show that this is so, and a more modern version of the idea can be found in the work of the American social philosopher George Herbert Mead. Mead's thesis is that the individual

> ... becomes a self in so far as he can take the attitude of another and act towards himself as others act. (Mead 1967, 171).

Being a self is reflexively structured: I myself relate to myself. According to Mead, we are not endowed with this structure from birth – rather, it evolves in the course of social interaction. And, as the quotation indicates, it is shaped through our adopting the same attitude towards ourselves as others have towards us. The formation of the self as a social structure involves two elements – namely, in part the individual's relation to the particular other, and in part a relation to what Mead calls the generalized other.

Through a relation to the concrete other, the self adheres to communication. As a behaviourist, Mead sees linguistic communication as such as an extension of the communicative processes we find in animals. Even in animals he finds the characteristic feature of one animal's "gesture" eliciting a response from another, and the former adapting its behaviour to this response. This feature takes on a reflexive aspect in humans, inasmuch as when the individual speaks or makes a communicative gesture, he or she relates to his or her own utterance from the perspective of the recipient, letting the recipient's anticipated reaction to a degree determine the form of the communicative act. Thus relation to self and meaning are inextricably intertwined: meaning simply rests on the fact that the individual can relate to an utterance in the same way as the other. The reflexive structure is linked to the identity and intersubjectivity of meaning.

To illustrate what it means to say that the individual relates to the generalized other, Mead makes a comparison with games. To participate in a game, we must assume a role, which means that we regard ourselves from the perspective of other players, or rather: from the perspective representative of the game as a social activity. It is as a self, relating to the generalized other, that the individual can act as a member of a community or society. When Mead captures this line of thought in the words "the organized community or social group ... gives to the individual his unity of self" (154) it might be given a communitarian gloss in the sense that the individual's identity and commitment to norms is always determined by the specific community he/she is actually a member of. This is not, however, Mead's view. What makes it possible for the individual to distance him/herself from his/her own community is that he/she is able, so to speak, to carry the generalization of the other beyond any specific social group and to adopt a perspective belonging to a "higher community", which ultimately comprises the whole of humanity.

To distance oneself in this way from a given community is now the same as to "*speak with the voice of reason to himself*" (167, my emphasis). The ability to see oneself from the perspective of the generalized other is, in other words, tantamount to the adoption of a rational attitude. In Mead, reflexive rationality becomes first and foremost a form of practical reason: it is the ability to see ourselves from the standpoint of the generalized other that enables us to enter into social cooperation.

As we have seen, constituent to modern conceptions of practical reason is the idea that the individual be able to look at him/herself from an impartial point of view, subjecting his/her own acts to a test of universalizability. The importance of Mead's theory lies in his showing that this aspect of practical reason has a social genesis, and that it comprises two distinct elements: the ability to put oneself in another's place and view oneself from that vantage point – and the ability to apply the perspective of the generalized other which is universal only in the final analysis.

Rational choice

If, against the backdrop of the foregoing brief survey (which naturally does not claim to be more than that) we return to the status of practical reason today, we may conclude that the tradition deriving from Hume, Bentham, and Smith still holds considerable sway. By this I mean that the conception of neutral means-end rationality still plays a central role in many of the social sciences' endeavours to describe social phenomena. The contemporary term for this kind of rationality is "rational choice". It has been summarized as follows:

> ... rational choice theorists normally agree on an instrumental conception of individual rationality according to which individuals are assumed to maximize their expected utility in formally predictable ways. When used empirically the further assumption is usually made that rationality is homogeneous across studied individuals (Green and Shapiro 1994, 17).

In this theory, rationality is not the impelling force behind human action. Instead, the primary motivational factor is the desire for maximal utility. Rationality enters into the matter when the individual is confronted with several options. In such cases rationality takes the form of specific requirements with regard to consistency, of which the most important are that (i) options must be ranked so that two outcomes are either equal or the one superior to the other, and (ii) preferential rankings are transitive (if A is better than B, and B is better than C, then A is better than C).

Rational choice theory represents the ascendancy of the tendency, introduced by Bentham, to conceive of social action in economic categories. The model is also used to describe political agency:

> ... The main goal of every party is the winning of elections. Thus all its actions are aimed at maximizing votes, and it treats policies as merely means to this end. (Downs, 1957, 35).

The dominance of the theory bears witness to a conception of practical reason which is the diametrical opposite of Aristotle's: whereas he subordinated both political and economic action to ethical reason (phronesis), economic rationality is now regarded as the rational norm for all social agency.

However, it must not be thought that rational choice theory is considered immune to criticism, even in the social sciences.[7] For instance, an economist such as Amartya Sen criticizes the narrow concept of reason presupposed by rational choice theory. He considers it a testament to the total loss of the link that economics once had with ethics, in both Aristotle and Adam Smith. His criticism is directed at the idea of self-interest being the sole motivation for rational action.

On the other hand, he does not contest the existence of self-interested rationality:

> it would be extraordinary if self-interest were not to play quite a major part in a great many decisions, and indeed normal economic transactions would break down if self-interest played no substantial part at all in our choices ... (Sen 1988, 19)

---

7. See, e.g., the chapter on rational choice in Root, 1993.

It is difficult to disagree with him here. It requires no great power of observation to ascertain that self-interested rationality is a thoroughly pervasive factor in modern society. That reason should be a spent force is no more than wishful thinking. The pressing question is, however, whether once self-interested rationality is discounted, morality amounts to no more than various irrational impulses – or whether it makes sense to speak of moral reason. If the second alternative is correct, the question that needs to be addressed next is whether moral reason is compatible with self-interested rationality.

MORAL REASON

Moral reason has many exponents in philosophy. And, understandably enough, they often tend to argue for Kant's conception of practical reason in a modified version. Kant's belief in reason that builds on indefeasible principles, making a "Letzbegründung" possible, finds few defenders today – at least, not in the form in which Kant conceived it. Instead we find the principle of universalizability grounded by reference to linguistic phenomena. For instance, R.M. Hare, a leading present-day utilitarian, justifies the principle on the basis of the key ethical term "must". But two other forms of moral Kantianism will be addressed in the following.

Habermas

Jürgen Habermas's theory of communicative reason and communicative acts is to a large extent an extension of Weber's theory of rationalization. Habermas draws attention to the fact that Weber sites the locus of rationalization in the structure of consciousness. Habermas sites it in language. This is also true of moral rationality. Habermas's conception of Kantian universalizability has it that the morally right action is that which everyone ought to perform in comparable situations, and so all must, in principle, be agreed on what the right act is. Habermas formulates the impartiality requirement in what he terms the axiom of universalizability U:

A norm is valid when the consequences and side effects for the individual resulting from everyone complying with it can be accepted by all affected parties ... (Habermas 1983, 75f. My translation.)

According to Habermas, U can be justified through its expressing premises essential to our being able to discuss on the whole. He terms this justification 'universal pragmatism', and it is here his conception of communication enters the picture. "Pragmatic" is a term used in diverse theories of language, often in conjunction with "semantic" and "syntactic". Semantics has to do with the meaning of words and sentences, in many instances with the relation between language and the world; syntax has to do with the correct form of sentences. Pragmatics (from the Greek *prattein*: do; the verb linked to the term praxis) is concerned with the fact that language is something we use: linguistic expressions are used by a speaker in order to produce an effect on one or more auditors. The pragmatic aspect of language has been the focus of the so-called theory of speech acts. In this context the expression "performative" has been introduced, to identify that feature of language in virtue of which we perform specific acts when we use language. If, for example, a doctor says to a patient "I'll come again tomorrow", he thereby issues a promise.

Taking his point of departure in speech act theory, Habermas elaborates his theory of communicative[8] acts, so-called. Its antithesis is strategic agency, which is result- or success-orientated. When I act strategically I follow an egocentric utility calculus, which is to say, I consider how I best bring about the results I want for myself. Naturally, I can also take others into account, but doing so amounts merely to letting what I expect their acts to be figure in my own utility calculus. Strategic agency corresponds roughly to what Weber terms means-end rationality, and today is called rational choice. Communicative agency, by contrast, takes others into account by aiming at understanding (Verständigung) in the sense of agreement, consensus. This form of agency is communicative, in that it presupposes a certain kind of speech. According to Habermas there are three types of speech acts: (i) those that say something about the state of the world; (ii) those that say something about valid social norms, and (iii) those that say something about the speaker. What the three types of speech act have in common is that, in performing them, we assert their validity. But they each have their own kind of validity, namely truth, correctness, and honesty, respectively. Communicative agency is a form of agency based on our recognition of the claims to validity connected to our speech acts.

---

8. The theory is presented in the major two-volume work, Habermas 1981. A brief summary is given in Habermas 1983, 144-152.

Moral discourse is the distinctive form of communication required when agreement on valid norms fails to prevail. Discourse is the procedure that is to ensure the realization of the universalizability principle. Discourse is governed by the discourse-ethical axiom D:

> A norm may only lay claim to validity when all who are affected by it reach agreement about its force, qua participants in practical discourse. (Habermas 1983, 75)

Discourse ethics is a form of ethical rationalism, in that Habermas assumes that a reasoned justification can be given for the validity of norms.[9] And rationalism is grounded in his conception of linguistic communication. He makes the assumption that persons as language-users are rational, in the sense that they are able to distinguish between the respective claims to validity of types of speech acts, and that speech consists in judging whether or not these claims to validity are fulfilled. An important presupposition for discourse ethics is that the language we converse in is sufficiently systematic in structure to allow all forms of interchange to be identified as belonging to one or other of the three categories of speech act. A point on which Weber, according to Habermas, is unclear, is whether universal rationality criteria exist. Habermas holds that there are such criteria. He recognizes that pluralism prevails in respect to the content of values. But he contends that there exist universal structural features of "moderne Lebenswelten überhaupt". (255)

Rawls

The most impressive attempt to defend practical reason as understood in the Kantian tradition is to be found in *John Rawls*. He is particularly relevant for the problem I formulated above, because one of his main

---

9. Habermas takes the view that moral discourse can and should be institutionalized in society. In the light of discourse ethics one might claim that the role of ethical advisory committees is to represent the various parties to an ethical disagreement and through "unforced dialogue" work towards a consensus on the matter in hand. Even if one does not agree that discourse grounds the validity of norms – as I do not – that does not prevent one from recognizing the usefulness of the conversational format in dealing with ethical disagreement.

interests is to develop a theory of practical reason in which ethics and the rationality of self-interest are united.

When dealing with Rawls' views on practical reason one should be aware of the fact that he is influenced by V.v.O.Quine in his views on the formation of scientific theory. Quine rejects "absolutist" views on scientific knowledge, such as those found in Kant and in logical positivism. His own views on science are holistic, in the sense that according to him, knowledge forms an interconnected unity, encompassing both everyday knowledge and abstract theories. And no part of our "web of belief" is definitive; adjustments can take place both in its empirical periphery and its theoretical interior. It is this view on rational knowledge that Rawls transfers to his understanding of ethical justification as it is expressed in the concept of *reflective equilibrium*. The equilibrium in question exists between our considered concrete moral judgements on the one hand, and general ethical principles on the other. The rationality that can be said to be implicit in reflective equilibrium is, however, not specific to practical reason.

The considered moral judgements we make in our capacity as citizens, on the other hand, i.e. as political beings, imply according to Rawls an understanding of ourselves as rational in a practical sense. It is this rationality that is reconstructed in the theoretical thought-experiment concerning the original social contract.[10] In the first version, Rawls presupposes that the contract partners are rational agents, more or less in the sense of the theory of rational choice. Only, they do not follow their own interests by aiming at maximum utility; their interests are directed towards goods of a different kind. I will return to this later. But Rawls still presupposes that the partners act according to well known strategies of rational decision making. As they are placed behind a "veil of ignorance", and thus do not know what their position in the future social system will be, they will follow a maximin strategy; i.e. they will choose those political-ethical principles which will give them the best conditions, should they be placed among the worst-off. Applied to the prin-

---

10. Rawls expressly claims that "justice as fairness" can be seen as a *Kantian* theory. He rejects the view that the crucial feature of Kant's ethical theory should be the principle of universalisation – in my opinion rightly so. The decisive feature, according to Rawls, is the idea of autonomy: a person can be said to act autonomously "when the principles of his action are chosen by him as a free and equal rational being" (Rawls 1972, 252).

ciples of justice, this strategy leads to the choice of the two well known ones of equal rights to freedom and the difference principle. The latter is a principle of distributive justice. According to Rawls, the object of just distribution is not utility or the pleasure of preference fulfilment, but basic social goods. One of the characteristic features of his theory of justice is that he regards the good as a rational concept (cf. the title of chapter VII in Rawls 1972: "Goodness as Rationality"). In this, he is up against a dominant tendency in modern understanding of the good. As we have seen, ever since Hobbes, the question of the good – the goal of human quests – has not been regarded as an issue for reason, but rather for the emotions, the passions, and the preferences. Reason was only supposed to relate to non-rationally set goals in the second place, e.g. by calculating the maximum net sum, or by choosing rationally between different options for preference-fulfilment.

Rawls defines goodness as rationality by extending the concept of decision rationality so that it covers *life plans*, although adding what he calls "deliberate rationality". Without going into detail, it should be mentioned that Rawls defines the good as "what a person having a rational life plan prefers". From this definition he arrives at the concept of basic social goods as a kind of minimal condition. It comprises, among other things, freedom, social opportunities, income, wealth and self-respect (Rawls 1972, 433). This is a minimal or "thin" definition of the good in the sense that, irrespective of what the individual regards as a good life, the goods mentioned are necessary prerequisites for realising such a life. Rawls' claim is thus that at least one essential aspect of the idea of the good life is rationally determined and hence universal. As already mentioned, the basic social goods are the "items" to be justly distributed according to the difference principle, stating that social and economic inequalities are only just if they are of the greatest advantage to the worst off, and if there is equality of opportunity.

The political ethics of *A Theory of Justice* is thus centred around a concept of practical reason that is a modification of rational choice theory. In the revised version of *Political Liberalism,* Rawls expressly makes it clear that he originally regarded the theory "as part of the theory of rational decision" (Rawls 1993, 53). This he now regards as erroneous, which is why he introduces a concept of practical reason distinct from "rationality", viz. "reasonableness".

Reasonableness is, so to speak, the reason of political ethics. According to modern tradition, the aim of the political order is not the achievement of common, given and pre-ordered goods. The political order is regarded instead as a framework of social cooperation among citizens who regard each other as free and equal. This way of regarding each other and the common life is in itself a manifestation of what Rawls means by "reasonableness". Furthermore, this concept implies being ready to propose principles of fair cooperation and the willingness to obey them. A crucial tenet of the theory is that social cooperation takes place for the sake of mutual advantage, and therefore involves rationality besides reasonableness: the individual seeks his/her own good, but (s)he is able to follow rules that ensure that the goods are achieved in a way that is fair to all. A political ethics which unites the two forms of reason in this way is, according to Rawls, different from both altruism and pure self-interest.

An important aspect of the new version of Rawls' justice as fairness is his recognition of *pluralism* as, not an anomaly, but rather an essential feature of liberal societies. He accepts an idea forwarded by communitarians, viz. that human beings are not exclusively rational or reasonable citizens, but also members of various communities. In the latter role they adhere to what Rawls calls "comprehensive doctrines", i.e. "Weltanschauungen". They, too, can be reasonable, namely if they allow their adherents to play the role of citizens in the manner described. In other words, a Weltanschauung is reasonable if it makes it possible for its adherents to have two elements in their view of reality: a publicly recognized political ethics (containing a principle of social justice) – and a comprehensive Weltanschauung. Among reasonable doctrines there will be what Rawls calls *overlapping consensus* about the principles of political ethics.

All in all, then, practical reason in Rawls' political liberalism can be regarded as a complementary entity, consisting of both rationality and reasonableness. The former is to be understood along the lines of rational choice theory, whereas the latter is the ability to act from an impartial point of view: to take part in social cooperation as a member on equal terms with others and in mutual recognition.

CONCLUSION

If one attempts to attain an overview of the many-faceted picture contemporary discussions of rationality present, one might arguably contend that Max Weber formulates the core problem: Does human rational agency consist merely in the choice of ends in accordance with preferences and the most effective pursuit of these – or is there in addition a form of reason which governs acts on the basis of values or other ethically normative concepts? In Habermas's discourse ethics and Rawls's theory of justice we have two proponents for answering the last part of the question in the affirmative. Both proposals are reformulations of Kantian rational ethics. The next question that presents itself is whether they necessarily carry a commitment to Kant's ethical anthropocentrism, or if it is possible to integrate concern for the environment and the natural order into a coherent theory of morally rational agency. Thus the complex concept of practical rationality is extended to cover contemporary environmental problems.

REFERENCES

Aristotle. *The Nichomachean Ethics. With an English Translation by H.Rackham.* Cambridge, London 1982.

Bentham, J. (1996). *An Introduction to the Principles of Morals and Legislation. An Authoritative Edition by J.H.Burns and H.L.A.Hart with a New Introduction by F.Rosen.* Oxford.

Downs, A. (1957). *An Economic Theory of Democracy.* New York.

Green, D.P. and Shapiro, I. (1994). *Pathologies of Rational Choice Theory: A Critique of Applications in Political Science.* New Haven

Habermas, J. (1983). *Moralbewusstsein und kommunikatives Handeln.* Frankfurt aM.

Habermas, J. (1981). *Theorie des kommunikativen Handelns. Band 1. Handlungsrationalität und gesellschaftliche Rationalisierung.* Frankfurt aM.

Henrich, D. (1976). "Die Grundstruktur der modernen Philosophie", i: Ebeling, H. (Hg.), *Subjektivität und Selbsterhaltung. Beiträge zur Diagnose der Moderne.* Frankfurt/M.

Höffe, O. (1976). "Einführung in die Wissenschaftstheorie der zweiten Analytik", i: Aristotle, *Lehre vom Beweis oder zweite Analytik (Organon IV).* Hamburg.

Hobbes, Th. (1983). *Leviathan. Edited and abridged with an Introduction by John Plamenatz.* Glasgow.

Hobbes, Th. (1991). *Man and Citizen (De Homine and De Cive). Edited with an Introduction by Bernard Gert.* Indianapolis/Cambridge.

Hume, D. (1978). *A Treatise of Human Nature. Edited, with an Analytical Index, by L.A.Selby-Bigge*. Oxford.

Hume, D. (1987). *An Enquiry Concerning the Principles of Morals. Edited, and with an Introduction, by J.B.Schniewind.* Indianapolis.

Kant, I. (1785). *Grundlegung zur Metaphysik der Sitten*. Riga.

Kant, I. (1788). *Critik der practischen Vernuft.* Riga.

MacIntyre, A. (1988). *Whose Justice? Which Rationality?* London.

Mead, G.H. (1967). *Mind, Self, and Society*. Chicago: The University of Chicago Press.

Mill, J.St. (1985). *Utilitarianism. On Liberty. Essay on Bentham. Edited with an Introduction by Mary Warnock*. Glasgow.

Rawls, J. (1972). *A Thory of Justice*. Oxford.

Rawls, J. (1993). *Political Liberalism*. New York:Columbia University Press.

Root, M. (1993). *Philosophy of Social Science*. Oxford.

Sen, A. (1988). *On Ethics and Economics*. Oxford.

Smith, A. (1994). *An Inquiry into the Nature and Causes of the Wealth of Nations. Edited, with an Introduction, Notes, Marginal Summary, and Enlarged Index by Edwin Cannan*. New York.

CHAPTER 2

Rationality and the Reconciliation of the DSP with the NEP

William E. Kilbourne and Suzanne C. Beckmann

THE ENVIRONMENTAL CRISIS

SINCE THE LATE 60's and early 1970's, the impact of business and consumer practices on the environment has received considerable attention in the public, political, and scientific debate. While interest has waxed and waned during the thirty years since its inception, there is currently a resurgence of interest in various dimensions of the "environmental crisis." Most of the research to date has focused on different aspects of the crisis and its various manifestations. Specific environmentally related forms of behaviours, for instance, have frequently been examined as researchers tried to precipitate environmentally friendly consumption behaviour through recycling, energy conservation, or the purchasing of "green" products. This was generally addressed through the assessment and transformation of attitudes, more specifically, environmental concern. In this approach, it was implicit that if individuals were environmentally concerned, they would alter their consumer behaviour in environmentally responsible ways.

This perspective was expanded more recently to examine the role of values as antecedents to environmental attitudes. It was argued that for environmentally related attitudes to be transformed, it was necessary to examine the value systems that gave them form and direction. Several researchers have begun to approach the problem in this way, and have demonstrated that values do play an important role in the process (McCarty & Schrum, 1994; Grunert & Juhl, 1995). There is, however, an acknowledged limitation in this approach as well. And that is that values themselves are supposed to be the function of cultural beliefs and institutions in which the individual is embedded (Stern, Dietz, &

Guagnano, 1995). This leads us to a perspective on the environmental crisis which is expanded even further, since it leads to an examination of the cultural frame of reference – or paradigms – rather than specific values, attitudes, or behaviours (Kilbourne & Beckmann, 1998).

PARADIGMS

The necessity of examining the environmental crisis as a crisis of paradigms was argued for by Dunlap and Van Liere (1978) who coined the term 'new environmental paradigm' (NEP) and contrasted it to the prevailing paradigm of Western industrial societies, referred to as the dominant social paradigm (DSP). The DSP was engendered during the Enlightenment, and has informed both scientific and social analysis since that time. Milbrath (1984) defines the DSP as "... the values, metaphysical beliefs, institutions, habits, etc. that collectively provide social lenses through which individuals and groups interpret their social world" (p. 7). Kuhn (1970) refers to a paradigm as a "disciplinary matrix", consisting of symbolic generalisations readily accepted by members of a community, belief in models of the relationship between objects of interest, and values regarding conduct within the paradigm. Cotgrove (1982) elaborates, suggesting that a paradigm is dominant, not because it is held by the majority of people in a society, but because it is held by dominant groups, who use it to legitimise and justify prevailing institutions. It becomes the justification for social and political action by the group and, as such, functions as ideology.

When interpreting and evaluating from within a paradigm, we examine existing conditions or beliefs and infer from them that the particular view of the world they engender "is" the world that "ought" to be. Hence, we reinforce the *status quo* and commit the naturalistic fallacy (Moore, 1903). This is the belief that no other outcome could have ensued than that which did ensue. We then conclude that whatever is, ought to be. This is precisely the error of the classical liberals, who accepted existing institutions, such as the distribution of property, as natural (Myrdal, 1952). But it is just those institutions that require justification. To say, for example, that economic growth solves social ills, when we should say it ought to solve them, justifies the growth ideology and condemns us to live with the problems of growth, leaving the institutions legitimating growth unexamined.

Within a paradigm, the usual approach to examining ultimate ends is to take conventional priorities, such as economic growth and technological development, as given, and determine from them the ultimate end that is consistent with and legitimates the *status quo*. Thus, if material prosperity is consistent with economic growth and technological progress, we will define it as the ultimate end. We then observe that the economy is growing and technology is advancing, and conclude that we are on the right path to the chosen end. In establishing intermediate ends, the criterion for their prioritisation is our ability to satisfy them with the preferred intermediate means that again are established within the DSP. This suggests that the means and ends of a society are not independent of each other, and are integrated into the social institutions of the DSP (Daly, 1991). Consequently, as suggested by Dunlap and Van Liere (1978), the environmental crisis is not one of environmental symptoms such as pollution or energy depletion, but of paradigms. While ameliorating the symptoms is a critical aspect of the remedy, it is fallacious to assume that eliminating pollution eliminates the crisis. The symptoms will simply manifest themselves in another time, place, or form if the problems are not addressed at the fundamental level of institutions and values, i.e., the DSP.

Since very little of the research to date has been at the level of paradigms, we might suspect that little progress has been made regarding the root causes of the environmental crisis. We should expect that some of the symptoms have been alleviated through technological developments, but that the basic elements of the crisis remain. This is the conclusion drawn by Smith (1995), who refers to this as the paradox of environmental policy. It is paradoxical because we seem to know both the short and long term solutions, but consistently fail to implement policy to effect relevant change. With regard to the current state of US environmental policy, Dowie (1995) argues that it has been shaped by the technological, political, and economic ideologies of the DSP. Consequently, the environmental movement must be reinvented if it is to be effective. This requires that it be a product of a new biospheric paradigm, which he refers to as the fourth wave of environmental activism. The question we must then ask is why there has been so little apparent success when so many express great concern for the state of the environment. It is argued here that this observation results from the unique origin and character of the crisis. We will now briefly examine the unique nature of the origins and then turn to an examination of the character of the environmental crisis.

THE UNIQUE ORIGIN OF THE CRISIS

Mesarovic and Pestel (1974) suggest that, unlike crises of the past, which have had negative origins, the current crisis is unique in that the factors precipitating it are perceived as positive. It is the product of successful intentions, rather than failures, as in the past. As such, it is extremely problematic, since the solution requires mobilizing the forces of change to combat a process largely considered to be "good" (Davies & Mauch, 1977). Here, the "good" is defined as such since it is one of the established ends within the DSP, i.e., higher levels of consumption. Hirsch (1976) also challenges the "goodness" of economic growth *per se*, suggesting that it "... undermines its social foundations" (p. 175). This recalls the earlier prognosis of capitalist decline by Schumpeter (1942), who argued that it would be capitalism's success that "... undermines the social institutions that protect it" (p. 61). In fact, he argued that capitalism contains the seeds of its own destruction, and this could well be what we are now beginning to experience as a crisis. The crisis is compounded by the fact that the decision criteria of industrial organisations, political institutions, and private households are predominantly short term and individualistic, ignoring both communitarian needs and the future.

Gladwin, Newbury, and Reiskin (1996) suggest that this mindset is characteristic of Northern elite cultures, and is contrary to sustainability. They argue that the "unsustainable mind" promulgates biases stemming from obsolete assumptions about how the world works, rationalisation of environmental deterioration, reliance on ideology, and inability to deal cognitively with systemic complexity. These factors combine to militate against holistic, communitarian, non-mechanistic thought. This unsustainable thinking characterises much of Western industrial societies' values, norms, beliefs, and attitudes.

The important dimensions of this problem are the cultural values associated with contemporary industrial society. Pirages and Erlich (1974) concur that the crisis stems from what were considered the strengths of the industrial social order. The dominant values held by the majority in Western societies are still the ones associated with these strengths, such as those relating to individualism, achievement, and private property. While the strength with which the values of the DSP are held may vary from country to country, the direction of beliefs is relatively consistent. Because environmental problems are predominantly value-related problems, technological solutions will not work in the long run. Yet such solutions are the mainstay of contemporary analysis, and are encapsulated within the theory of ecological modernisation, suggesting that the institutions of these societies need not be critically examined (Kilbourne, McDonagh, & Prothero, 1997).

THE CHARACTER OF THE CRISIS

It is evident from society's many conflicts on the subject that there is no consensus about what constitutes rationality. Because cultural life has been fragmented into a collection of relatively disjointed compartments rather than an integrated whole, rationality may be variously argued to rest in the pursuit of self-interest (economic), the role of the citizen (political), efficiency (technological), or the pursuit of the greatest good for society (utilitarian ethics), and each of these dimensions might well have different forms of rationality. As a consequence of this fragmentation, contemporary political structures have few public forums through which to explore conflict (Thurow, 1980; Dryzek, 1996). Disagreements are not acknowledged as such because the compartmentalized issues are decontextualized and viewed ahistorically. The incompatibility of beliefs within the paradigm is not examined, and discourse is reduced to an agreement to disagree. This is characteristic of debates between environmentalists and business, for example. Each challenges the premises of the other, and the debate is always inconclusive.

One solution to this is to distance one's self from the established social relations within which conceptions of the good, responsibilities, and interests have been framed. This enables a neutral decision, but fails to account for different conceptions of rationality. In addition, establishing a procedure predicated on neutrality in the first place leads to a certain set of acceptable conclusions. Specifically, debate in Western industrial societies is typically predicated on atomised individualism and procedural neutrality, each of which must first be justified. If they are not, the premise then contains the conclusion. So separating one's self from a context appears to be illusory, since once the conditions of discourse have been settled, a particular form of rationality follows. Kuhn (1970) states of this difficulty: "The alternative is not some hypothetical 'fixed' vision, but vision through another paradigm …" (p. 128). But the historically and socially bound contexts of an established paradigm are inescapable. The result is not rationality, as suggested, but discourse, involving counter-assertions based on incompatible sets of premises.

Liberal, Enlightenment rationality has become the dominant mode of rationality in the modern world (Rifkin 1980). It is considered by its proponents to set the standard against which the peculiarities of other rationalities are judged, and it is the only acceptable alternative for circumventing these cultural peculiarities in the application of reason. In Figure 1 we present this idea in a simple framework with two competing rationalities, or rational fields, subsumed under a superordinate rational field. The superordinate rational field is equivalent to rationality *as such,* and is marked by dashes to indicate that this field does not exist.

Each rational field – and there can be many – is characterised as a set of objects of interest within the field but not necessarily found in other fields. Each field also has characteristic institutions that determine how the objects of interest are to be interpreted and combined to yield new objects. These are the + and = in Field 1 and the & and => in Field 2, effectively representing Kuhn's symbolic generalizations, models, and values. An individual in one field has little or no knowledge of the objects or institutions in the other field, but each field has its unique form of rationality, which is the focus of discourse, and is understood by its members. The two fields in this case are non-overlapping sets, indicating that the proponents of each have no communicative capacity with the other.

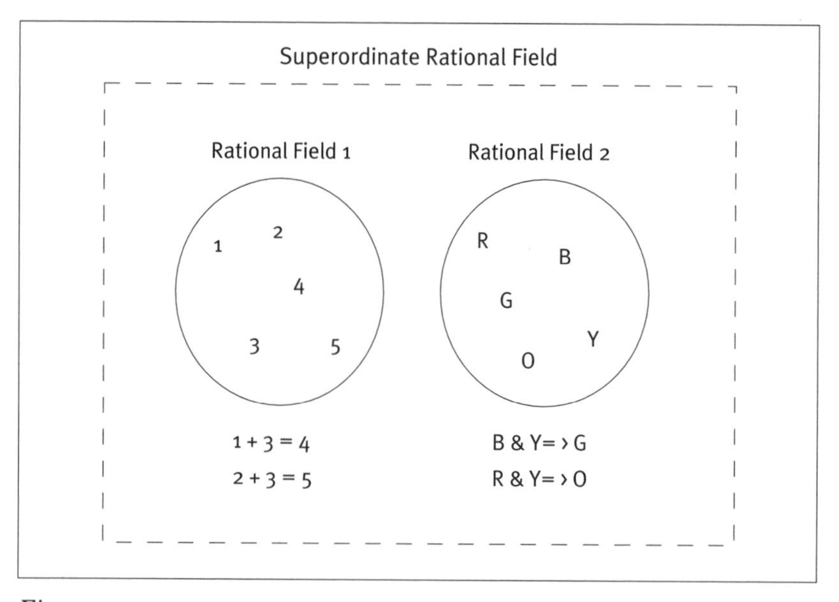

Figure 1

In the historical competition for paradigm-dominance, Enlightenment rationality has excluded other alternatives, even though this standard of rationality has failed to fulfil its promise of universality. Any alternative will be presented as a contending view, which cannot be justified against Enlightenment standards. Hence all alternatives will be rejected when critique is framed in the idiom of the *status quo*. As a particular conception of rationality becomes institutionalised, it is inured, rendering its constitution opaque, even to its proponents. With increasing ascendancy over the standards of rationality that preceded it, it becomes not only the best alternative, but also the only one. The standards of rationality of the *status quo* have been shown to work within that paradigm by better overcoming limitations and remedying defects than those that preceded them (Kuhn, 1970). However, they are not considered paradigms as such until they begin to fail or are confronted by another paradigm. From this we can conclude that a rational person is only rational within a particular paradigm, while not even recognising the paradigm as such. In addition, the requirements for justification are different in different paradigms, and these competing doctrines can only be understood from within the historical context in which they evolved. This suggests that there is not a single rationality, but many rationalities (MacIntyre, 1988).

That there are many rationalities does not imply that differences cannot be rationally resolved, but seems to require a post-Enlightenment superordinate rationality. Before this can happen, though, it must be recognised that modern liberalism, predicated on the demise of paradigm-based rationalities, has become the modern paradigm.

To understand the environmental crisis as a crisis of paradigms requires that the debate be framed in these terms, and not as an isolated problem of, for instance, technology or economics. In this sense, we can substitute DSP for Rational Field 1, and NEP for Rational Field 2 in Figure 1. In doing so, we have transformed the debate from one of public policy alternatives to one of alternative paradigms. It is argued here that the DSP has evolved from Enlightenment liberalism, and has as its key elements expansive market economics (Kassiola, 1990; Schmookler, 1993), technological rationality (Habermas, 1970; Ehrenfeld, 1978), and the politics of procedural neutrality (Sandel, 1996; O'Neill, 1993). It stands in opposition to the NEP, which is predicated on steady-state economics (Daly, 1991; Sagoff, 1988), appropriate technology (Schumacher, 1973; Winner, 1986), and deliberative democracy (Dryzek, 1996; Gutmann & Thompson, 1996). The distinction between a conflict of policy alternatives and a conflict of paradigms cannot be overstated, as it is within this distinction that the environmental crisis must be examined. Steady-state economics cannot be evaluated as failed growth economics; "green" political thought is not new laws or a new parliament, and appropriate technology is not fewer applications of larger technologies. The DSP and the NEP represent opposing world views, and failure to examine the environmental crisis as a crisis of paradigms will result in a reconfirmation of liberalism as the paradigm of choice, so that the crisis will continue to worsen, despite minor successes achieved through new legislation or applications of technological development.

LIBERALISM AS A PARADIGM

The characteristic feature of paradigms is the existence of debate within them and their capacity of transforming themselves at points of conflict. Thus when problems of incommensurability and untranslatability appear within a paradigm, its adherents find a rational way of reconciling the differences from within the paradigm, so that the anomalous becomes the expected (Kuhn, 1970). Recognising the inadequacy of intellectual resources within a paradigm is essential to its growth and development, and is a sign of its maturity. The failure of modern liberalism is not so much that it fails to answer its own questions adequately, but that it has not resulted in the universality of rationality, as its project demanded: it has failed to become the paradigm beyond paradigms.

Meanwhile, the environmental debate proceeds apace, with neither position giving ground. Because the liberal paradigm is the dominant one, it demands that the debate be framed in its own idiom, i.e., the context of the liberal market as the mediator not only of physical resources, but of intellectual ones as well. Thus the economic dimension, while acknowledging the severity of environmental problems, argues that the market should be allowed to function, and, when it fails to do so efficiently, market-like mechanisms (pollution permits, bubble policies, etc.) should be used. If this fails, then technological solutions can be developed that will erase the tracks of market inefficiencies. If both fail, more stringent legislation and better enforcement of existing legislation are required. In none of these cases is the modern, liberal paradigm itself called into question. Proponents of the NEP argue that none of these approaches represent a long-term solution, since they do not address the root causes of the crisis, only its symptoms. Instead they call for sweeping structural change at the institutional level, challenging the DSP and its institutions. Since this is necessarily unacceptable to the *status quo,* which controls the level of debate, the conflict between the two positions intensifies.

The conflict between the DSP and the NEP is evidence that the liberal promise of a superordinate rationality has failed, since if the liberal paradigm had succeeded, the conflict ought to have been resolved, or at least not appear immutable. The questions raised in the NEP should have been resolved as a normal part of the functioning of the DSP. Their resolution, as suggested earlier, appears no closer now than three decades ago, as the essential tensions between the two paradigms remain. The atomistic anthropocentrism of the DSP is incommensurable with the holistic ecocentrism of the NEP, and the proponents continue to talk past each other in a barrage of assertions and counter-assertions of irrationality emanating from competing rationalities. Neither can reconcile the differences of the other from within its paradigm. There is no ground on which to stand outside one's position, and no language to provide neutral accounts of the givens of either paradigm.

Because of this limitation, claims of the rational superiority of one paradigm over another are not justified, since a common point of observation which forms an adequate basis for rational evaluation cannot be found in either. Within any paradigm, the choice of action requires resorting to a particular theory or theory-presupposing scheme, and the concepts so generated will be in accord with the theory articulated at the outset. Ecocentric premises result in ecocentric conclusions about reality, and anthropocentric premises result in a confirmation of the DSP. If, for example, one chose to demonstrate that advancing technology was beneficial to society, one would choose an example demonstrating that economic growth and material well being followed in the wake of such advances. Since economic growth and material well being are mainstays of the DSP and form the definition of progress, the result would confirm the theory. There is no neutral ground for choosing between rival theories, and any such attempt would result in vindication of the theory, since one of the elements of the theory relates to how the example will be framed and described. Because this is established by the theorist, any appeal to example will vindicate the theory and reject a competing paradigm (MacIntyre, 1988). Hence, if proponents of the DSP test the adequacy of a competing rationality, such as that of the

NEP, they will frame the challenge in terms of free markets and their ability to produce material prosperity. It will then be clear that the outcome within the DSP is superior to the outcome under the NEP. Conversely, if proponents of the NEP issue the challenge, they will frame it in terms of a different allocative mechanism and a different criterion of success, such as economic justice. The NEP will be demonstrated to be superior in this test. So this suggests that whoever frames the challenge will always prove to be correct. Since the DSP is "dominant," its proponents are in a position to frame all challenges and determine the outcome.

The project of the modern, liberal, individualist society is precisely to develop a social order emancipating the individual from the contingencies of any paradigm. The appeal is to universal norms, independent of tradition, and in the history of this search, its attainment appears illusory. In the failure to secure a position based on principles of shared rationality, liberalism has been transformed into a paradigm (MacIntyre, 1988). Liberal principles allow a disparate group of individuals to coexist in the pursuit of different conceptions of the good life. All are free to choose their own conception, provided that it does not hinder another's similar pursuit. The political process is to remain neutral toward the individual's concept, and attempts to institutionalise any one concept are proscribed (Rawls, 1993). The difficulty is, of course, that liberal individualism holds its own conception of the good, and is actively engaged in imposing it politically, economically, and socially (Sandel, 1996). The market mentality is the cornerstone of the DSP solution to the environmental crisis, and as a consequence, its proposals must remain within its own constructions for the resolution of environmental anomalies. This frames the problem as a market failure rather than the failure of markets to exist (Sagoff, 1988). Within the market failure construction, only a limited set of alternatives consistent with liberal theory is possible, none of which suggest a challenge to the paradigm itself. Once again, the NEP flatly rejects this approach, arguing that the problem, the market mentality, cannot at the same time be the solution. If atomised individuals pursuing their own self-interest in free markets leads to the tragedy of the commons, how, they ask, can more of the same be a solution?

As a final contradiction, the DSP also increases intolerance toward rival conceptions of the good, negating the purpose for which the liberal project was engendered. Like other political and economic conflicts, rival conceptions of the good engendered within the NEP must pass the market test. But we have already established that the market will always over-produce private goods and fail to provide adequate public goods such as environmental amenities, goods, confirming that what we have been doing is what we ought to be doing and serving as further protection of the *status quo*, i.e., we have allowed the market to function, and material prosperity has increased. Since material prosperity is the primary criterion of success within the DSP, the market vindicates itself.

THE LIBERAL MARKET TEST

Nonliberal theories only gain exposure through the expressions of preferences in the market. But even this is through excision, not voice. I can express a contrary choice of the good by not choosing an existing one, but *why* I choose is not considered (Dryzek, 1987). There is no voice except the *sotto voce* of the liberal market, and it is limited to those with sufficient bargaining power to make even this voice heard (O'Neill, 1993). This expresses the notion that in liberal society, there is no superordinate good, but separate and distinct spheres in which a good might be pursued. Because of this diversity of goods, no overall ordering of goods for the purpose of collective social choice is possible (Arrow, 1951), and this becomes the norm in liberal society. Rawls (1971) argues that to be otherwise is irrational, since the self is heterogeneous. Within the liberal framework, only the rules are explicit, yielding a procedural rationality with the goal of enabling individuals to exercise their preferences. While Dryzek (1987) defends ecological rationality as prior to competing forms, the cogency of any argument is irrelevant, since the only valid test is the market. Ultimately, the test fails, since the criteria of acceptance are established within the liberal market itself, and include, for example, economic growth, material well being, and maximisation of individual choice, all anathema to ecological rationality. This is the essence of liberal society (Kassiola, 1990; Sandel, 1996), and the means by which it defends its own principles. Distributive justice is, at least in theory, enacted as a constraint on the rules, to insure fair participation in pursuit of preferences, even by the disadvantaged. But even this remains a peripheral criterion of effectiveness in the liberal market, which invokes efficiency in the form of Pareto optimality as the tribunal of economic rationality.

The transformation of thought and action precipitated during the Enlightenment (Hirschman, 1977) elevated the role of preferences, making them compatible with liberal economics and politics (Mirowski, 1989). The confluence of liberal economics and politics with the psychology of the liberal individual defined a new reasoning individual, who was characterised as engaging in rationality *as such*. But such individuals are really exercising a form of practical reasoning specific to their own paradigm. The critical aspects of this practical rationality in contemporary Western societies are the ordering of preferences in the market, and how preferences are to be summed up to yield a just society. However, this entails only social justice, and ignores the elements of ecological justice. But even here, the outcome depends upon the premises for the constitution of justice.

Because conflicting conclusions always ensue from this debate, the only rational approach is to first decide which sets of premises are true: e.g., does the welfare of the present generation take precedence over that of future generations? Should ecological conditions take precedence over economic growth? The answer to these questions requires that we establish a set of premises at the outset and then develop procedures that will yield the desired result. But in liberal theory all such claims are equally (in)valid, because no superordinate good can be justified. Gutmann and Thompson (1996) argue that, in such cases, deliberation should precede choice, and that if we are to confront moral disagreement, "... we must not check our deliberative dispositions at the door to the public forums" (p. 38). Without such forums, the debate is inconclusive, and discourse is replaced by persuasion or power. Rational argument then recedes and is replaced by the expression of preferences in the market.

Reverting to the liberal market, however, presupposes that the conditions of procedural justice, i.e., impersonal market mechanisms as the arbiter of resource claims, were themselves the product of rational debate in a state approaching the Rawlsian original position. In this condition, individuals with a shared rationality develop principles of justice with no knowledge of their position in society or of prevailing conceptions of the good (Rawls, 1993). But the debate over conceptions of the good remains perpetually inconclusive, because there is no longer a public forum in which the debate can occur under truly participatory conditions. Still, achieving a consensus on conceptions of the good remains the ostensible goal, providing the illusion that progress is being made toward its attainment. Mirowski (1988, 1989), however, questions whether, in the development of economic rationality within the DSP, any debate ever really took place.

This takes us from discourse (political economy) to the market (neoclassical economics) and back to discourse (deliberative democracy) for the justification of the rules by which preferences are pursued, i.e., what are the procedures of procedural justice and what in them accounts for ecological rationality. These rules must be shown to be egalitarian and ecologically sound if proponents of the NEP are to be made to show allegiance to them. It is here that the DSP again falters. The debate on effective procedures, like the others, remains inconclusive, so we revert once again to preferences within the liberal market.

With preferences playing the crucial role in liberal society, however, the power to determine the alternatives from which preferences are determined becomes the pivotal position in society. This role has been arrogated by an elite which controls not only the objects of preference themselves (Parenti, 1978), but also how the alternatives are symbolically represented (Featherstone, 1991). Both the objects and the images are constructed in the idiom of the DSP as representations of material well being and individual freedom. It appears then, that the appeal of liberalism is not to a paradigm-independent rationality, but to one of many different possible paradigms.

TRANSFORMATION OF PARADIGMS

Thus far, it has been argued that the rationality of the DSP cannot be critiqued from within the NEP, since such a critique requires a perspective independent of either paradigm. No such vantage point is in evidence. While, at certain stages of development, a paradigm might provide rational justification for its own concepts, there is no independent, superordinate rationality by which to examine others. Thus, when confronted with competing paradigms, as in the environmental debate, there is no rational basis for choosing between them. There is no rationality as such. Resorting to the market for resolution takes us no further, since it presupposes the conditions that were to be justified by the debate.

Rational inquiry begins within a paradigm as a product of prevailing beliefs and institutions in that paradigm at a particular time. These are the givens of any paradigm. However, historically accepted truths can become problematic as anomalies become evident. Such a condition may develop, for example, as two divergent paradigms come into contact with one another, as shown in Figure 3. With such changes, some of the older beliefs are maintained, but the new forms come into increasing conflict with them, and become more rationally justifiable as the inadequacies of thought in the old paradigm are revealed. When environmental problems such as pollution were first confronted, they represented such an anomaly within the economic rationality of the DSP. They were then defined as externalities and were not the subject of much debate in the normal conduct of economics. As their ineluctable character became apparent, they commanded greater attention, assuming the position of a subfield (environmental economics). The inadequacy of thought in formulating the problem is revealed when the problems remain intransigent even as more resources are allocated to their solution within the constraints of the old paradigm.

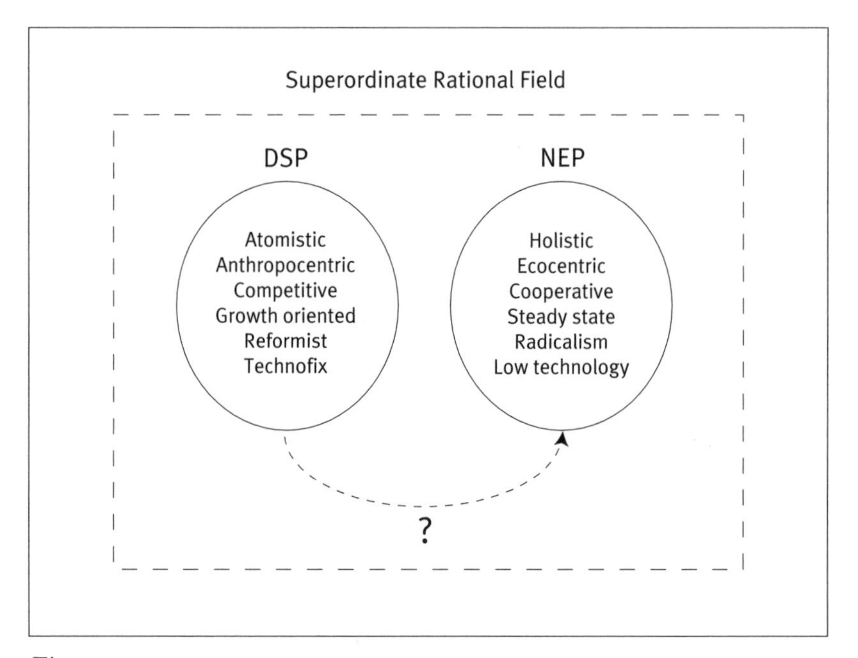

Figure 2

These discrepancies must be reconciled in new forms that appear to be immune to the environmental crisis that created them. While forms of rationality vary from paradigm to paradigm, commonalities will become evident, but forms of argument and the requirements of success in dialectical reasoning are yet to be established. For instance, it is becoming apparent that the intellectual resources of the DSP are inadequate in alleviating environmental degradation.

MERGING THE DSP AND THE NEP

If it can be shown that some of the principles of the NEP are more coherent than their DSP counterparts, they will be incorporated into the old paradigm. In this dialectical process those that are shown to be superior will be sustained, but these are neither self-evident nor self-justifying; this they become, as their superiority in solving the current problem is established. As the number of accepted beliefs diminishes, an epistemological crisis occurs, requiring new concepts and theories to furnish a solution to the crisis and an explanation for why the incoherence occurred in the first place: e.g, holism must be demonstrably better than atomism as a mode of thought. This must be done while some of the original character of the DSP is maintained, in which case, a new perspective on the humanity/nature relationship can be established which maintains some of the original flavour of atomism. These new positions will not be derivable from the old, but represent innovation and imagination. There is no logical way to arrive at the conclusion that ecocentrism is superior to anthropocentrism from the concepts of either paradigm. Each takes its own position as a first principle from which others are derived, but one cannot be derived from the other.

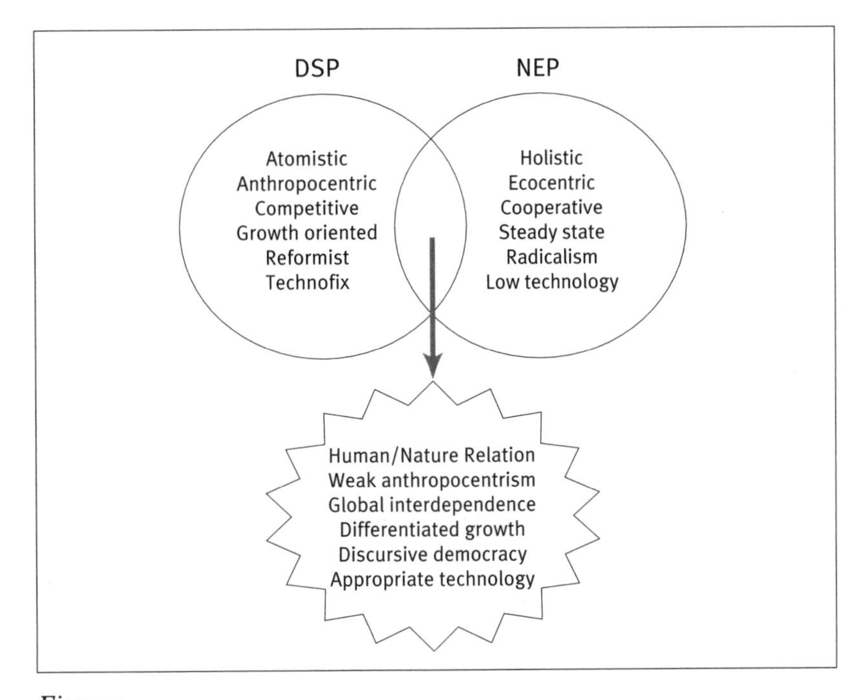

Figure 3

In passing through such a crisis, those who believe in the old paradigm can redefine their beliefs in a more insightful way, incorporating elements of the new paradigm and maintaining continuity with those aspects of the old paradigm which have survived the crisis. This process must be accepted as a condition of any paradigm change, since an epistemological crisis may occur at any stage of its development. When incoherence or the inability to solve its standard problems appear, it must be recognized that new conceptual resources are required to restore coherence (Kuhn, 1970). If the elements of the NEP are acknowledged to provide a better explanation for why a crisis exists and how to solve it, then the established claims of the DSP have been defeated. If its proponents do not acknowledge this, it is possible that the old paradigm will be defeated by its own inadequacies. This is the position of those less optimistic critics who suggest that only ecological disaster unequivocally attributed to the inadequacies of the DSP will be sufficient for a transformation of paradigms.

Hence, the argument suggesting that no paradigm can enter into rational debate with another is unfounded. Under normal circumstances, that may be true. But in times of epistemological crisis, it is not necessarily so. Instead, the determining factor will be how a paradigm responds to its crisis.

CONTESTED RATIONALITIES

It has been argued here that the peculiar rationality of the DSP is embedded in a particular view of life and the place of that life in nature. This world view is then embodied in specific types of social relations. Thus individuals are enmeshed in their historical circumstances, making them a part of the problems that they are confronted with. This is not peripheral to the problem, but its very essence. The interests and needs of individuals do not operate independently of presuppositions about their place in life, but are derived from them. It is not our interests and needs which have motivated and guided the development of the DSP, but a mutually reinforcing reflexive process. As a result, in the coming epistemological crisis, competing forms of rationality, such as the NEP, will make their bids for superiority and must be able to demonstrate the ideological character of the DSP while themselves remaining reality-based.

As competing paradigms make claims, they do so within the realm of liberalism, the prevailing paradigm, and must adhere to the rules of that paradigm. In doing so, debate about a paradigm is pre-empted and reformulated into policy debate in the idiom of liberalism. We may debate the form a policy may take, but not the fundamental tenants of liberalism itself. The problem for proponents of the NEP is the creation of a forum in which the terms of the debate do not predetermine the outcome (Dryzek, 1996). When engaged in the debate on policy within the liberal forum, the standards of rationality are assumed to be available and agreed upon by individuals, regardless of the paradigm from which they come. Once this claim to universality is rejected, it becomes clear that each conception of rationality or policy depends on one's paradigm. But the problems are not the same to everyone, even within a paradigm: what a problem is, and how it is addressed and resolved are historically and socially dependent.

A truly intellectual engagement requires individuals to make inquiry both within their own paradigm and between rival paradigms. The latter requires that the person essentially become a member of the competing paradigm in order to understand and see the world through different eyes. While most do not live at the extremes of paradigms, neither are they free from them. While the dominant liberal, individualist form is unquestioningly accepted, the individual also draws from a variety of paradigm-generated resources, resulting in inconsistent beliefs, which remain untested. This protective compartmentalisation of the self produces individuals who tolerate different rationalities. When new paradigms are introduced, individuals must systematically test elements of the new against the old. This requires learning the idiom of each in order to describe the other. One can then determine which of the rival paradigms best develops the narrative of their own existence. Discourse in terms of rationality *as such*, beyond paradigms, is in effect impossible. To pursue such a discourse effectively precludes outside voices from being heard, and serves a protective function for the paradigm. Establishing the conditions of debate without context or history insures that no debate ensues (MacIntyre, 1988).

REFERENCES

Arrow, Kenneth J. (1951). *Social choice and individual values*. New Haven, CT: Yale University Press.

Cotgrove, Stephen (1982). *Catastrophe or cornucopia: The environment, politics and the future*. New York: John Wiley & Sons.

Daly, Herman E. (1991). *Steady-state economics*. Washington, D.C.: Island Press.

Davies, Joan, & Mauch, Samuel (1977). Strategies for societal development. In: Meadows, Dennis L. (Ed.), *Alternatives to growth – I: A search for sustainable futures*, pp. 217-242. Cambridge, MA: Ballinger.

Dowie, Mark (1995). *Losing ground*. Cambridge, MA: The MIT Press.

Dryzek, John S. (1987). *Rational ecology: Environment and political economy*. Oxford: Basil Blackwell.

Dryzek, John S. (1996). *Democracy in capitalist times: Ideals, limits, and struggles*. Oxford: Oxford University Press.

Dunlap, Riley, & van Liere, Kent (1978). The 'New Environmental Paradigm.' *Journal of Environmental Education*, 9(4), 10-19.

Ehrenfeld, David (1978). *The arrogance of humanism*. New York: Oxford University Press.

Featherstone, Mike (1991). *Consumer culture & postmodernism*. London: Sage Publications.

Gladwin, Thomas N., Newburry, William E., & Reiskin, Edward D. (1996). Why is the Northern elite mind biased against community, the environment and a sustainable future? Keynote speech presented at the 3rd conference of the Nordic Business Environmental Management Network, Aarhus,Denmark.

Grunert, Suzanne C. & Juhl, Hans Jørn (1995). Values, environmental attitudes, and buying of organic foods. *Journal of Economic Psychology*, 16(1), 39-62.

Gutmann, Amy, & Thompson, Dennis (1996). *Democracy and disagreement*. Cambridge, MA: The Belknap Press of Harvard University.

Habermas, Jürgen (1970). *Toward a rational society: Student protest, science, and society*. London: Heinemann.

Hirsch, Fred (1976). *The social limits to growth*. Cambridge, MA: Harvard University Press.

Hirschman, Albert O. (1977). *The passions and the interests: Political arguments for capitalism before its triumph*. Princeton: Princeton University Press.

Kassiola, Joel (1990). *The death of industrial civilization*. Albany, N.Y.: State University of New York Press.

Kilbourne, William E., & Beckmann, Suzanne C. (1998). Review and critical assessment of research on marketing and the environment. *Journal of Marketing Management*, 14(6), 513-532.

Kilbourne, William E., McDonagh, Pierre, & Prothero, Andrea (1997). Can macromarketing replace the Dominant Social Paradigm? Sustainable consumption and the quality of life. *Journal of Macromarketing*, 17(1), 4-24.

Kuhn, Thomas (1970). *The structure of scientific revolutions*. Chicago: Chicago University Press.

MacIntyre, Alasdair (1988). *Whose justice? Whose rationality?* Notre Dame. In: University of Notre Dame Press.

McCarty, John A., & Schrum, L.J. (1994). The recycling of solid wastes: Personal values, value orientations, and attitudes about recycling as antecedents of recycling behaviour. *Journal of Business Research*, 30, 53-62.

Mesarovic, Mihailo, & Pestel, Eduard (1974). *Mankind at the turning point*. 2nd report to the Club of Rome. New York: E.P. Dutton.

Milbrath, Lester (1984). *Environmentalists: Vanguard for a New Society*. Albany, N.Y.: State University of New York Press.

Mirowksi, Philip (1988). *Against mechanism*. Lanham, MD: Rowman and Littlefield.

Mirowski, Philip (1989). *More heat than light*. Boston: Cambridge University Press.

Moore, George Edward (1903). *Principia ethica*. Reprinted 1971, Stuttgart: Reclam.

Myrdal, Gunnar (1952). *The political element in the development of economic theory*. London: Transaction Publishers.

O'Neill, John (1993). *Ecology, policy and politics*. London: Routledge.

Parenti, Michael (1978). *Power and the powerless*. New York: St. Martin's Press.

Pirages, Dennis, & Ehrlich, Paul R. (1974). *Ark II: Social response to environmental imperatives*. San Francisco, CA: W.H. Freeman.

Rawls, John (1971). *A theory of justice*. Cambridge, MA: Harvard University Press.

Rawls, John (1993). *Political liberalism*. New York: Columbia University Press.

Rifkin, Jeremy (1980). *Entropy*. New York: Bantam Books.

Sagoff, Mark (1988). Some problems with environmental economics. *Environmental Ethics*, 10, 53-74.

Sandel, M. J. (1996). *Democracy's discontent*. Cambridge, MA: Harvard University Press.

Schmookler, Andrew Bard (1993). *The illusion of choice. How the market economy shapes our destiny*. Albany, NY: State University of New York Press.

Schumacher, E. Fritz (1973). *Small is beautiful: Economics as if people mattered*. New York: Harper and Row.

Schumpeter, Joseph A. (1942). *Capitalism, socialism and democracy*. New York: Harper and Row.

Smith, Zachary E. (1995). *The environmental policy paradox*. Englewood Cliffs, NJ: Prentice Hall.

Stern, Paul C., Dietz, Thomas, & Guagnano, Gregory A. (1995). The New Ecological Paradigm in social-psychological context. *Environment and Behaviour*, 27(6), 723-743.

Thurow, Lester C. (1980). *The Zero-Sum Society: Distribution and the possibilities for economic change*. New York: Basic Books.

Winner, Langdon (1986). *The Whale and the Reactor: A search for limits in an Age of high technology*. Chicago: University of Chicago Press.

Part II

CHAPTER 3

Rationality, Institutions and Environmental Governance

Martin Enevoldsen

INTRODUCTION

THIS ARTICLE EXAMINES the conceptual and theoretical basis on which policy-makers construct instrumental means to reduce pollution. Amidst the enormous variety and complexity of system dynamics and actor characteristics, the individual rationality of the actors stands out as the central ontological principle on which modern environmental regulation has been built. First, environmental problems are believed to be caused by short-sighted pursuit of material advantages and the neglect of broader social interests. Second, the regulatory means advanced to solve the problems have been overwhelmingly oriented towards prohibiting or weakening the individual's impulse to disregard harmful environmental side effects. Thus, the introduction of legislation containing sanctions and/or rewards is largely an attempt to make it rational for the polluters to behave in accordance with the long-term interests of society.

The problem, however, is that traditional command-and-control regulation has often failed to deliver the environmental improvements it was supposed to (Jänicke, 1990a; Weale, 1992). Although the failures can to some extent be explained by inadequate scientific understanding and limited governmental capacity to address the interaction and problem displacements between complex environmental issues, they certainly also have to do with the regulators' inadequate perception of polluter motives and behaviour. The reliance on rational models, assuming on the one hand a frictionless implementation of environmental regulation,, and on the other hand perfectly informed private utility-maximising, has proven too simple. It has failed to recognize the uncertainty surrounding environmental problems and the related search for cleaner technologies. Policy makers, administrators, and private subjects (firms and citizens, who carry out the bulk of pollution abatement activities) do not have complete information on the environmental problems with which they are dealing. Their choices are to a large extent guided by institutional procedures, e.g., the path of earlier choices, norms, routines, and rules of thumb. Moreover, further environmental progress is premised on probabilistic chances of realising essentially unknown technical innovation. These chances depend on learning, creativity, and luck. It should therefore come as no surprise that environmental policies seldom work in the way predicted by simplistic top-down models.

What seems to be needed is a theoretical basis more capable of grasping the complex interaction between individual motivations and rationality, on the one hand, and social institutional factors on the other. This chapter is an attempt to approach the question through a critical discussion of a selection of theories addressing the relation between the means of environmental governance and the choices of the polluting actors. In the following sections, a rational approach (section 2) and an institutionalist approach (section 3) to environmental governance are discerned. Although the major theories of the two approaches have advanced our understanding of the subject in various ways, the analysis will point out their respective limits in accounting for the conditions of successful environmental policy. In the final section 4, I will draw some conclusions on what the rational and institutionalist approaches have to offer with respect to the study and practise of environmental governance at the beginning of the 21st century. Which problems have been solved by the theories, and which remain? What normative implications can be drawn regarding the use of environmental policy instruments?

THE RATIONAL APPROACH TO ENVIRONMENTAL GOVERNANCE

In the following sections, I will focus on the strands in rational choice theory[11] that explicitly address the causal relation between pollution

11. Rational choice theory has roots in the realist and utilitarian traditions, especially the lines of thought advanced by Thomas Hobbes and Jeremy Bentham. Being less interested in the continental project of developing a

control instruments and the polluter's choices, and that have used the insights to formulate normative prescriptions for rational environmental governance. Two candidates have been selected. One is the collective action perspective on environmental protection, which has focused on the need for either direct regulation or privatisation in order to prevent individual inclinations toward environmentally harmful action from being realised at the expense of collective welfare. The second is the Pigouvian externality theory and its proposal to regulate environmental problems by way of economic instruments that make the polluters internalise the externality costs they impose on society. I will present the main propositions of each perspective and discuss some internal and external critique that may be raised against them before concluding on their merits and limitations in accounting for effective environmental governance. But first I will present the core assumptions of the rational choice model that forms the basis for both perspectives.

One common characteristic of all rational choice theories is methodological individualism. The individual is the basic unit of analysis. Methodological individualism requires that every macro-theory on social relations be founded on a micro-theory of individual behaviour. Accordingly, the aim of rational choice studies of politics is "... to explain political strategies and, so far as possible, economic and political outcomes by showing how these strategies result from the self-defined interests of actors" (Katzenstein, 1984: 34). Another characteristic of rational choice theories is the common stock of underlying rationality assumptions. Rational choice theory assumes that individuals behave according to the following four assumptions; the first and second relating to their motivations, and the third and fourth relating to their capabilities:

Utility maximising
Stable, exogenous preferences
Consistent preference orderings
Unlimited cognitive capabilities

The assumption that people maximize individual utility is the causal driver in rational choice theory. It describes the dispositions and inclin-

---

→ holistic understanding of human reason and rationality, rational choice theory attempts – on a simple deductive basis – to explain individual behavior in social affairs.

ations of economic man; what he *will* do when faced with any choice of alternatives. He will weigh the costs and benefits of each alternative and choose the one that maximizes his own utility as defined by his preferences. "The basic behavioural postulate of public choice, as for economics, is that man is an egoistic, rational utility maximizer." (Mueller, 1989: 2). But does the inclination to maximize personal utility necessarily imply that individuals are selfish? For example, Geoffrey Brennan has argued that rational choice theory ought to be devoid of substantial content. Hence, it is a mistake to include the assumption of self-interested behaviour (Brennan, 1990).

The justification for viewing selfishness as an essential part of rational choice can be traced back to Thomas Hobbes' ideas on man's desire for self-preservation, and to some extent also to Darwinian arguments.[12] It appears that the assumption of self-interest has prevailed in economic theory ever since Adam Smith (Arrow and Hahn, 1971), although it has not always been openly stated and properly defined (cf. Sen, 1982). In contrast, public choice theories have been very explicit about the subject.[13] However, some rational choice theorists have argued that altruism as well as egoism can serve as a forceful motivation, since the well-being (or misery) of others may enter the utility calculations of the individual (Taylor, 1987; Margolis, 1982). But in the rational choice perspective, altruism is generally seen as an exceptional rather than a general source of motivation (cf. Mansbridge, 1995).

The second assumption relates to the standard for personal utility, that is, the underlying preferences. Preferences simply express what a person wants, or prefers, at a given point in time. The preferences of the individual are assumed to be given once and for all. They are exogenous to the choice situations facing the person, and they are unaffected by his actual choices (see e.g. Arrow, 1963; Becker and Stigler, 1977). Rational

---

12. Along with other animals, man is presumed to be endowed with a selfish gene that enhances his prospect for survival and reproduction. However, the Darwinian argument is double-edged. The notions of group and kin selection, which emphasize the indispensability of collective orientations as a means for survival, is a clear case against a unique status for selfishness (cf. Margolis, 1982: 26-35; Elster, 1979: 141ff).
13. For further details, cf. Downs (1957: 3-28), Buchanan and Tullock (1962: 17-39), Riker and Ordeshook, (1973: 8-37), and the critical reviews in Margolis (1982) and Lewin (1991).

choice theory attempts neither to explain where preferences come from nor how they are shaped. The third and fourth assumptions postulate the existence of certain individual capabilities that enable the maximising of utility in a consistent, thorough manner.

Consistent preference orderings refer to the individual's capability of ordering the utility of alternative goods in a complete, transitive and continuous manner (Elster, 1979; Heap et al., 1992). *Completeness* requires that with any pair of options, the individual should always be able to express a preference for one of them, or failing this, indifference. In other words, any two bundles of goods can always be compared and ranked in terms of individual utility. *Transitivity* requires that the preferences of the individual are consistent. Formally, if a rational person prefers option A to option B, and option B to option C, she will also prefer option A to option C. *Continuity* basically means that, from the point of view of individual utility, all goods are tradable, or "everything has a price" (Borch, 1968: 22). Thus, given a bundle made up of two different goods, it will always be possible – by reducing the proportion of one good and increasing the proportion of the other – to define another bundle which is just as useful for the individual.

Finally, in order to maximize utility, the rational decision-maker is assumed capable of obtaining information on all aspects of all possible choice alternatives, and capable of performing calculations demonstrating their relative advantage (cf. Simon, 1957). Hence, instrumental rationality boils down to the twin assumption of perfect information and unlimited computational abilities (cf. North, 1990; 17-26; Hammond, 1996: 117). There are, however, degrees of difference; not every rational choice theory adheres to the strict version. Unlike mainstream neo-classical economics, for example, game theory does not assume the players to have perfect information (Rasmussen, 1994). Perfect information implies not only that the players know all past choices made by the other players, but also an ability to foresee future choices. Being a reaction to the so-called "paradox of perfect foresight" (Morgenstern, 1935 [1976]; Knudsen, 1997: 84ff), game theory explicitly recognizes the uncertainty stemming from the unobservability of future moves. Moreover, game theory recognizes the existence of games where it is impossible for one player to observe the past choices of another player. In contrast to perfect information, game theory typically makes the more modest assumption of *complete information.*

The collective action perspective on environmental protection

The collective action perspective on environmental protection mainly belongs to the game-theoretical strand of rational choice theory. Originally, the authors in this perspective used the insights provided by Mancur Olson (Olson, 1965) to explain the existence of environmental problems as a logical consequence of rational choices. Based on these analyses, solutions were proposed of how to deal with the problems through specific legal instruments at the disposal of nation states (cf. the critical summary in Ostrom, 1990: ch.1).

According to the collective action perspective, environmental protection (pollution) is classified as a more or less perfect collective good (evil) (R. Hardin, 1982; Sandler, 1992). In the absence of state intervention, an under-supply of collective goods is anticipated, due to private suppliers' difficulties in achieving compensation for the non-exclusive and non-rival benefits they bring to society. Since rational individuals consider only the private costs of a collective evil such as pollution, and not the costs imposed on the rest of the community, the risk is that society ends up with environmental damage making everyone worse off. In such a situation, individuals would be better off by limiting pollution on a mutual basis, but it is nevertheless unlikely that such collective action can be sustained on a voluntary basis, since each member has an incentive for a free ride, regardless of what the others do. The problem arises whenever the pay-off structure resembles a prisoners' dilemma game.

The established view of environmental problems as collective action dilemmas owes much to Hardin's famous essay "The tragedy of the commons" (G. Hardin, 1968). He considers an open-access pasture of limited size to which the herders from a near-by village bring their sheep to graze. By bringing another sheep, the herder obtains a private marginal benefit (e.g. higher yield of wool, meat or milk), whereas the marginal depletion he inflicts on the commons is a cost that will be shared by all herders. Hence, it may be rational for the individual herder to add livestock to the commons beyond the level where the new sheep impose a net loss on the community of herders. But if the other herders act in the same way, the accelerating depletion of the commons will make the profit of each herder shrink below the original level. The problem, however, is that even if the herder foresees the problem of over-grazing, he still has an incentive to add as many sheep as possible; his restraint alone would not make much difference to depletion, and he still has an interest in squeezing out a little extra benefit. "Therein lies the tragedy: Each man is locked into a system that compels him to increase his herd without limit – in a world that is limited" (G. Hardin, 1968: 1244). Inspired by this interpretation, the commons problem has been formally described and analysed as a prisoners' dilemma game (e.g. Dawes, 1973; R. Hardin, 1971; 1982; Schelling, 1978: 110ff).

The association of the commons problem with the prisoners' dilemma metaphor has been very effective. During the 1970s and 1980s, scholars, policy-makers and other interested observers formed the belief that appropriators of common-pool resources were unable to reach cooperative solutions on their own. The observers saw only two possible ways out of the dilemma (Ostrom, 1990: 8ff). Some recommended the use of coercive force by a Leviathan, i.e. state regulation prohibiting overuse of common pool resources. When a sufficiently large penalty is imposed on appropriators for exceeding the prescribed optimal appropriation quotas, it will no longer be profitable for them to overuse the resource – provided that the regulator is able to trace the violation and is willing to impose the penalty. The other recommended solution was privatisation of the commons. The idea is to hand out private property rights for each appropriator of part of the territory. Then each appropriator is playing a game against nature, where self-interest urges him not to overuse his own plot. Based on historical evidence from the United States, Libecap has argued that commons problems were better handled in those cases where some system of property rights existed (Libecap, 1989). However, the solution presupposes that the plot owners are able to keep others out, which can be quite costly in the form of fence building, mobilization of sanctions, etc. Apart from the procedural and ethical reservations with regard to the command-and-control and the privatisation solutions, there are several reasons for questioning the theoretical basis on which they have been formulated:

Ostrom has gathered a substantial amount of evidence on cases where communities of appropriators have managed to set up voluntary self-governing regimes characterized by stable appropriation and maintenance of common pool resources (Ostrom, 1990; Ostrom et al., 1994). The well-documented voluntary solutions to the commons problem contradict the rational choice prediction that collective action will fail. One possible explanation is that the prisoners' dilemma is a misleading description of the strategic situation facing the actors. Some authors have argued that the strategic incentives related to the provision of collective goods sometimes have the character of a chicken game, in which the contradiction between individual and collective rationality is less outspoken due to the strong incentive to avoid mutual defection (Taylor, 1987; Lipnowski and Maital, 1983). The chicken game analysis, in particular, has been proposed as an explanation of the commons problem in fisheries (Taylor and Ward, 1982). Fishery can only be exploited up to a critical level, beyond which there is a catastrophic collapse. Thus, there is a point beyond which it is not profitable to add another fishing vessel, even if the marginal costs of depletion are spread over the whole community, and late-comer appropriators may therefore choose to stay out.

However, another and more general reason why long-enduring, stable appropriation of common-pool resources can work out, even in the absence of state intervention, owes to the existence and development of social institutions – a largely neglected factor in the causal theory proposed by the collective action perspective. Typically, social norms on how to deal with the most pressing problems emerge as a result of the recurrent encounters and exchanges of communication between the appropriators. Schotter observes that:

> "… there is a learning process going on in which the players learn the type of behavior they can expect from each other and build up a set of commonly held norms of behavior. It is upon these commonly established norms that a social convention of behavior or institution is established that prescribes behavior for each participant in the conflict" (Schotter, 1981: 39).

Under favourable circumstances, well-designed institutions can help to transform non-cooperative exploitation of common pool resources into cooperative and sustainable appropriation, enforced by binding collective agreements (Ostrom, 1990). Ostrom therefore comes to the surprising conclusion that traditional policy prescriptions, recommending either state intervention or privatisation of common pool resources as the only way out of the problem, are unfounded (1990: 1-23). In many cases, but not all, the commons problem is handled more effectively by self-governance in the form of voluntary institutional arrangements among appropriators living in the community (Ostrom et al, 1992).

Finally, the theory on environmental choices proposed by the collective action perspective can be criticized for having a limited scope of relevance; after all, it has only been convincingly demonstrated for the commons problem and international negotiations on transboundary pollution problems (e.g. Hanley and Folmer, 1998). The tragedy of the commons is an environmental problem, in the sense that it refers to non-sustainable exploitation of renewable resources (e.g. deforestation, extinction of wild life and plant species, water drainage, over-fishing, etc.). Pollution abatement problems, on the other hand, concern the search for and innovation of environmentally benign technologies, including source-oriented cleaner technologies. The action perspectives of individual polluters differ fundamentally between the two classes of environmental problems. First, pollution abatement problems are more open to technological solutions, but also imply more uncertainty, due to the complexities and future contingencies associated with the employment and innovation of such solutions. Second, whereas restraint in the use of common pool resources approximates a pure collective good, technical pollution abatement may also entail considerable private benefits, e.g., conversion to cleaner technologies, leading to savings in production input.[14] Third, it is important to make a distinction between the interdependent local communities of resource appropriators, and the diffuse extensive communities composed by a mix of polluters and ordinary citizens. Polluters in extensive communities are normally not much worried about the diffuse damage inflicted on them by other polluters, as it is far outweighed by private benefits realised from their own polluting activities. Thus, in contrast to resource appropriators, most polluters face no dilemma in making decisions about their economic activities.[15]

The reservations mentioned above imply that problems of cleaner technology are not well apprehended by the traditional collective action perspective. Moreover, the collective action perspective has not been able to incorporate the kind of institutional factors found by empirical studies to be essential for stable, enduring resource appropriation. The problems have contributed to its declining importance as a descriptive

---

14. Pollution abatement is therefore a “joint product” (cf. Sandler, 1982) rather than a pure collective good.
15. Polluting firms will normally prefer that all firms, including their own, can carry on with their polluting activities without restrictions, rather than being relieved of the pollution impact of other firms at the cost of severe restrictions on their own economic activities. Since mutual pollution is pareto-optimal in the game among polluters, it follows that the pay-off structure is deadlock rather than prisoners’ dilemma. Accordingly, it is the citizens, not the polluters, who face a collective action dilemma.

and normative theory on which to base national environmental regulation (Dasgupta, 1990). As a result, the focus of the perspective has turned towards research on international negotiations regarding transboundary pollution problems (see e.g. Mäler, 1989; Hanley and Folmer, 1998). Here the application of non-cooperative game theory is much more straightforward, as there is no public authority, and very few social institutions, to enforce the environmental commitments. In the last section of the chapter, I will argue that the collective action perspective may also shed light on how to deal with some new environmental problems from the most recent history of national environmental regulation. One is the emergence of voluntary environmental agreements as an alternative to traditional regulation. Another is the pressing need for cooperative technology networks in order to further the innovation of cleaner technologies. But, as shown above, the collective action perspective has many flaws, and therefore needs to be held up against other perspectives.

The Pigouvian externality theory and environmental economics

In "The Economics of Welfare", Pigou showed that certain economic activities give rise to positive or negative externalities (e.g. pollution), which are not reflected in the self-interested behaviour of the economic agents. (Pigou, 1920 [1946]). The Pigouvian externality theory belongs to the neo-classical strand of rational choice theory. It treats the decision of each economic agent as a simple, isolated maximum problem. Thus, when making decisions on the character and amount of polluting activities, the economic agent chooses entirely on the basis of his own production function. Whether he operates in a competitive market or as a monopolist, price formation logic allows him to treat the other actors (including other polluters, competitors and consumers) as objects, i.e., as a set of stable parameters on which to base his decision. This is in contrast to the emphasis on strategical choice in collective action theory (and oligopoly theory), according to which each agent makes a choice on the basis of the choices he expects the other agents to make. Despite this fundamental difference, the collective action theory has borrowed the concept of externality costs from the Pigouvian tradition (Schelling, 1983). But whereas the strategic externality concept is suitable to problems of resource appropriation in small, interdependent communities, the parametric externality concept of Pigouvian theory appears more relevant in accounting for the diffuse market-wide consequences of effluent pollution.

Pigou was the first to show that divergences between marginal private utility and marginal social utility can be mitigated by centrally imposed taxes (Pigou, 1920 [1946]: 172-212). With respect to environmental problems, Pigou demonstrated that certain productive activities give rise to negative externalities in the form of pollution; these are not incorporated in the cost function of producers, but appear as a cost to others in society. Thus, there is a welfare loss whenever the marginal private benefits of a productive activity are less than the sum of private costs and net externality costs imposed on society. Social efficiency can, however, be restored by the state: the imposition on producers of charges (and/or bounties), which adequately reflect the externality costs (or benefits) of production, will encourage producers to adjust their output in line with the social optimum.

The Pigouvian tradition constitutes a cornerstone of the modern discipline of environmental economics. Environmental economics have shown that a socially efficient outcome can be achieved by means of a properly designed *effluent tax*. In order to find the correct tax level, the authorities must first estimate the socially efficient extent of pollution reduction. The efficient quantity appears exactly where the marginal social benefits of pollution reduction equal the combined[16] marginal abatement costs of all polluting firms. Hereafter, optimal pollution reduction can be achieved by imposing a tax on each unit of effluent pollution, forcing firms to reduce pollution up to the efficient amount, after which marginal abatement costs begin to exceed the tax level. This is the argument for the *static efficiency* of a correctly designed effluent tax. Environmental economists recognize, however, that precise estimation of the damage caused by pollution, on the one hand, and the sum of private abatement costs, on the other, is a formidable task. In reality, governments cannot live up to it, and the Pigouvian solution therefore has to be replaced by a proxy-solution, where the tax is set so as to achieve politically determined goals for total pollution, which do not necessarily represent the efficient level (Baumol and Oates, 1988: 159-176).

---

16. The combined marginal abatement costs are obtained by summing up the marginal private abatement costs of all polluting firms (Perman et al., 1996: 237).

As regards this second-best case, environmental economics theory has nevertheless shown that the effluent tax constitutes a *cost-effective* instrument for reducing pollution. Although there is no guarantee that the established amount of pollution reduction will be achieved by the tax level chosen, due to regulators' insufficient knowledge of enterprises true abatement costs, the resulting pollution reductions can still be achieved at the lowest possible costs. Moreover it can be proved that the tax instrument offers an improvement in cost-effectiveness, compared to a uniform regulatory standard. At any given tax level, each firm has an incentive to carry out only the degree of pollution reductions that are profitable, seen from its own particular cost structure. This means that pollution abatement will be carried out at those production sites where it is most profitable to do so. In contrast, uniform standards imply that some firms will be forced to carry out excessive pollution abatement given their particular cost structure, whereas others could have contributed even more at lower costs. Uniform standards thus imply a social welfare loss compared to effluent taxes.[17]

One of the potential advantages of green taxes is therefore that they do not presuppose nearly as much rational insight on behalf of the central authority as does command-and-control regulation. Using the price mechanism, regulation by means of green taxes leaves it to the polluters to decide where, when and how much pollution to abate. If command-and-control regulation were to match the cost-effectiveness of this approach, it would have to define individually optimal standards for each polluting enterprise, which again would require perfect information on the costs and technological opportunities of each firm. For green taxes to be effective, it is sufficient that the firms themselves have this information. Another major advantage of green taxes is that they provide dynamic incentives for the innovation of new cost-reducing pollution abatement technologies.

> "If compliance with an order is costly and if there is some choice of how to comply (what equipment or technique to use) then there will be an incentive for the source faced with the order to seek cheaper ways of complying in the long run. It is also true that for any particular source, an incentive system that puts a value on the discharge

---

17. Pearce and Turner (1990) offer a simple graphical illustration of the cost-effectiveness advantage of charges over standards.

> remaining after control will create a greater incentive to change than will a regulation specifying that same level of discharge" (Bohm and Russell, 1985: 417).

Although the dynamic efficiency of green taxes is not a well-established theoretical fact, recent contributions in the field of environmental economics have demonstrated that green taxes at least offer more incentive for technical innovation than direct regulation.[18]

While the assumption of a rational, omniscient regulator has to some extent been dispensed with by the proxy-solution proposed by modern environmental economics, the postulated cost-effectiveness of green taxes still requires that the firms behave according to the rational choice model. It is assumed that each firm has perfect information on how its cost structure will be affected by any relevant abatement technology, and that each firm is prepared to choose, and capable of choosing, the cost-minimising alternative on this basis. The Pigouvian theory, in other words, presupposes a frictionless market in which information is costless and abatement technology devoid of any rigidities, so as to enable utility maximising along neatly convex abatement cost curves.

The problems that green taxes have encountered in actual practice are due not only to improper institutional design (i.e. that the policy-makers ignored the economists' advice), but also to the friction associated with private market solutions to pollution abatement problems. Coase has described market friction in terms of transaction costs, including search and information costs, bargaining and decision costs, and policing and enforcement costs (Coase, 1988: 6ff). The transaction-cost barriers to optimum behaviour under green taxes are especially owing to the time-consuming search for optimal pollution abatement technology and the uncertainties related to technical innovation. But the administrative transaction costs of designing, negotiating and implementing regulatory instruments must also be taken into account before any conclusion can be made regarding the relative cost-effectiveness of green taxes vis-à-vis other forms of regulation (Coase, 1960; 1988: 157ff). Moreover, Coase demonstrated that a private agreement between the polluter and the victims of pollution may be as efficient as Pigouvian taxes, provided that it does not give rise to higher transaction costs.

---

18. See e.g. Downing and White (1986); Milliman and Prince (1989); Kemp 1997).

With respect to technical pollution abatement problems, there are good reasons to believe that Pigouvian tax-bounty instruments and tradable emission permits (the other instrument proposed by environmental economics) will be more efficient than command-and-control regulation. First, decentralization of abatement decisions under a price constraint (or quantity constraint in the case of tradable permits) supports flexible decision-making at the company level, and carries a dynamic incentive for technical innovation. Second, although the economic instruments are not immune to governmental deficiencies, they are nevertheless more robust than command-and-control regulation. But the fundamental discrepancies between the underlying causal theory and the complex, uncertain, institutionalised settings in which real-world actors operate remind us that we must be sceptical as to the merits of economic instruments. The bounded rationality of the actors generates additional uncertainty and transaction costs, and this in turn stimulates institutional arrangements to provide an orderly structure for cooperation and economic exchanges (North, 1990). However, the rational approach to environmental regulation has been unable to account for the ways in which contextual institutional factors influence the effectiveness of regulatory instruments.

THE INSTITUTIONALIST APPROACH TO ENVIRONMENTAL GOVERNANCE

To what extent do institutions have an impact on the effectiveness of environmental governance, and what institutions are particularly important in this respect? The natural place to seek the answers to these questions is in the institutionalist literature on environmental regulation. Two forms of institutionalism are relevant for this purpose. One is rational institutionalism, which analyses choices under uncertainty, from the perspective that the actors are boundedly rational and constrained by institutions. There have been only a few analyses of environmental governance in this perspective. The major contributions are those of Coase and Ostrom, which I used in the critique of the rational approach in the preceding sections. Another perspective is the broader neo-institutionalism, which more explicitly distances the rational choice model, especially the assumption of self-interested utility-maximizing. However, attempts to specify the institutional factors structuring environmental governance have never tended towards the extremist neo-institutionalist view that individuals are mindless role-conformers or powerless rule-followers. In fact, the theories under review in this part of the section are grounded on the observation that, more often than not, environmental regulation has not accomplished the results intended by the rule-makers.

Policy implementation theory

One source of inspiration to the institutionalist-oriented approach is Renate Mayntz's study of German air and water pollution control, carried out on behalf of the Federal German Council on Environmental Quality because of concern that the environmental legislation suffered from an implementation deficit (Mayntz et al., 1978). Mayntz concluded that mandatory environmental standards do not automatically result in the prescribed target group behaviour. Considerable effort must be invested in control and monitoring activities, as well as in the prosecution of violators, if regulations are to be effectively enforced. These control efforts by public agencies are subject to practical and economic limitations, and since the limits are often reached before the sanctions become sufficiently credible, many polluters choose to run the risk of not complying with the standards. However, Mayntz also saw the implementation deficit as a natural consequence of the bargaining processes between polluters, inspectors and local permit authorities. As many other studies have confirmed, it is quite typical that local environmental authorities choose a pragmatic response to companies that have difficulties in complying with centrally imposed standards (see e.g. Downing and Hanf, 1983). A cosy relationship may develop between the inspectors and the companies, and even if it does not, local authorities will often listen to good arguments for upholding a permit despite certain deviations from the standards.

Earlier studies of other policy fields had already shown that formal objectives of statutory legislation often went unrealised as a consequence of problems of administration and enforcement in the implementation process (Derthick, 1972; Pressman and Wildawsky, 1973; Bardach, 1974). The so-called "top-down" theories on political implementation focused on discovering the major causal variables leading programmes astray on their way down the implementation hierarchy. Top-down theories used these insights to formulate recommendations on how to improve goal achievement through better legal structuring of the critical downstream barriers in the implementation process.

Pressman and Wildawsky argued that the chance of successful goal achievement depends on the number of decision points and clearances throughout the implementation process. If we assume even a small probability that implementation will be halted or otherwise distorted at every point where joint action is required in order to move to the next link in realising the program, it follows that program success is highly improbable. With reference to this "complexity of joint action", Pressman and Wildawsky called for more realistic expectations, but also proposed a top-down system of control, communication, and resources to keep track of the individuals and organizations involved in the performance of implementation tasks.

Thus, from the beginning, implementation theory was concerned with searching out more rational ways of governing the implementation process (see e.g. Hood, 1976; Dunsire, 1978; Gunn, 1978). One of the most influential contributions in the top-down approach, and probably the most useful for public management purposes, was provided by Sabatier and Mazmanian (Sabatier and Mazmanian, 1979; 1983 [1989]). Based on comparative, instead of single-case analysis, Sabatier and Mazmanian proposed a general theoretical framework, which accounted to a greater extent than earlier studies for the statutory means by which to structure the implementation process in accordance with the legally mandated objectives. The legal structuring was seen as one of two other classes of variables – the tractability of the policy problem and broader political factors – on which successful implementation depends. As a result of the empirical analyses, six variables were identified as being necessary and (usually) sufficient conditions for effective implementation of the legal objectives (1983 [1989]: 41-42).

1. The formulation of clear and consistent legal objectives
2. The incorporation of an adequate causal theory
   The enabling legislation must incorporate a sound theory identifying the principal factors and causal linkages affecting policy objectives and it should give implementing officials sufficient jurisdiction over target groups and other points of leverage as a means of ascertaining the causal assumptions
3. A legal structuring of the implementation process so as to maximize compliance among implementing officials and target groups
   Limits on the number of veto points. Overcoming of local resistance through sanctions and incentives.
   Task assignment to sympathetic agencies characterized by hierarchic-

al integration, supportive decision rules, sufficient financial resources, and adequate access to supporters
4. Appointment of committed and skilful implementing officials
5. Mobilization of support from interest groups and sovereigns
6. Avoidance of socio-economic changes that undermine political support or causal theory

There are a number of problems with the top-down approach to implementation. First of all, reservation must be expressed as to the "implicit and unquestioned assumption that policymakers control the organizational, political, and technical processes that affect implementation" (Elmore, 1979: 603). It fails to take into account that policy outcomes are very often determined through social interactions between street-level bureaucrats, target groups, and other private associations with a more direct interest in, and better knowledge of, the actual problems than those on the top of the hierarchy (Hanf and Scharpf, 1978; Lipsky, 1979). Too much control of these decentred influences is neither practicable nor desirable. Even if better correspondence between the policy output and the legally mandated objective could be achieved by more control, there is no guarantee that the resulting policy outcome would be better for society.

As a reaction to the problems with the top-down approach, an alternative bottom-up approach developed, based especially on a number of works by European scholars at the Science Center of Berlin (Hjern et al., 1978; Hjern and Hull, 1982). Confronting what they saw as the myth of top-down control, the bottom-up scholars asserted that implementation studies should focus on the strategic interactions involved in local implementation structures. Moreover, the bottom-up approach questioned the evaluative standard of the top-down approach: "The bottom-up approach is essentially agnostic as to the evaluative yardsticks to be applied. It leaves the researcher free to apply the top's objectives as evaluative yardstick, or to apply those of any other actor" (Hanf, 1982: 171). The shifts in perspective led to the suggestion of an inductive, bottom-up method for reconstructing the local networks involved in implementation:

> "The methodological imperative in implementation research ... implies less a choosing than a discovering of the basic unit of analysis. Having defined the problem which is the focus of his analysis, the implementation researcher sets out to reconstruct what actors – with

what objectives, strategies and resources – are involved in the policy system" (Hjern and Hull, 1982: 110)

The advantage of this method is that: (a) it enables the researcher to discover decisive local factors which escape the attention of top-down approaches; (b) it provides a perspective for answering: what problems have been solved? (instead of: which official goals have been met?); (c) its applicability is not restricted to traditional regulatory policies. The local factors are also important from the point of view of policy-makers, and Elmore therefore recommended 'backward mapping' as a possible method of describing policies that have a greater chance of both being effective and meeting the needs of the local society (Elmore, 1979). However, the main problem with the bottom-up approach is its weakness in accounting for the relative theoretical significance of the numerous forces at play in local bargaining over implementation. In other words, it has a powerful and replicable method, but a weak theoretical framework (Sabatier, 1986: 35). As a result, the bottom-up approach has been quite successful in describing and mapping the implementation process, and in understanding the underlying policy problems from different perspectives, but much less successful in explaining why some regulatory policies came out better than others.

In addition to the above critique of implementation theory, there is, in my view, a more profound problem that questions its relevance as an approach to explaining the effectiveness of current environmental governance. Implementation theory has been explicitly developed to explain the implementation of command-and-control regulation (typically legally mandated standards enacted through a series of agency decisions and enforced by local authorities, usually with a substantial margin of discretion). With this kind of governance, implementation deficits are very likely to occur. The new kinds of indirect environmental regulation –e.g. green taxes and environmental agreements – do not present the same kind of implementation problems. Although it is correct that green tax legislation must also be implemented (emissions must be monitored, taxes must be collected etc.), one crucial difference is the absence of discretion. Tax levels are fixed in the legislation, and failure to collect or pay the full amount is a very risky criminal offence. Only if green taxes are combined with important regulatory provisions on tax exemptions, subsidies, etc., does the old implementation problem reappear. Another crucial difference is that, compared to direct regulation, implementation decisions are vastly more decentralized under indirect regulation; green taxes, for instance, leave all interesting decisions concerning abatement levels, technology choices, etc. to the firms involved. With respect to the implementation of voluntary environmental agreements, legal hierarchies and public authorities play an even more marginal role. After having negotiated the overall environmental targets with the private parties, the public authority has little to do except overseeing that the targets are complied with.

It is therefore no wonder that the top-down approach has lost significance as various forms of indirect regulation have begun to replace traditional regulatory policies. However, since direct regulation is still preferred for many types of environmental problems, the top-down approach clearly has some relevance in explaining possible implementation deficits. One the face of it, it would appear that the bottom-up approach, which does not presuppose a hierarchical process governed from the top, is much more suitable for analysing the implementation of indirect regulation. It can be argued that its inductive methodology is relevant for identifying the network actors contributing to the implementation of a decentralized, cooperative instrument such as voluntary agreements. Nevertheless, I doubt that the bottom-up approach, in its original formulation, will be able to cast much light on the factors influencing the policy outcomes of green taxes and environmental agreements. As already explained, bottom uppers have very broad views on the local policy process, and have not developed theoretical tools for integrating the many variables into clear causal explanations of policy outcomes. My final conclusion is therefore that traditional implementation theory does not offer much help in accounting for the effectiveness of indirect environmental regulation. This also applies to its newer, improved versions, e.g. the advocacy coalition theory (Sabatier and Jenkins-Smith, 1993; 1998) which constitutes a highly useful approach to explaining policy choice, policy learning and policy change. But through this latter specialization, the theory has also removed itself from the original question of instrument effectiveness that concerns us here.

The Berlin school: conditions for successful environmental policy

Compared to the rational approach to environmental governance, implementation theory has managed to breech the barrier between public rules and private incentives. It draws attention to bargaining processes and organizational arrangements with a biasing influence on policy implementation. However, implementation studies did not take an explicitly institutionalist perspective (most of them came before the rise of neo-institutionalism in political science, dating from around the publication of the seminal article by March and Olsen (1984). The implementation studies indicated the relevance of institutional factors through their focus on organizational arrangements and other constraints on political interests involved in implementation bargaining processes. But the relation between choice and institutional factors was never specified in the original studies. This stands in contrast to the more explicit institutionalist perspective of a subsequent approach that I will now discuss: the "Berlin school", in which Martin Jänicke's theory on ecological modernization capacity is recognized as the most notable contribution.

The Berlin school has affiliations with the bottom-up theories on implementation formulated at the Science Center of Berlin (see preceding section). The bottom-up ideas spread to environmental scholars at the Science Center, in particular Helmut Weidner –former colleague of Martin Jänicke at the Free University of Berlin, and his closest collaborator. Notwithstanding Jänicke's more structural approach, his work on state failure is clearly inspired by the bottom-up opposition to what it saw as the 'noble lie' of rational top-down governance. At the same time, his critical attitude was an echo of the broader debate among contemporary German sociologists and philosophers (Offe, Luhmann, Teubner, etc.) stressing the essential "ungovernability" of society (see e.g. Kooimann, 1993). Finally, Jänicke also took up ideas by the founder of ecological modernization theory, Joseph Huber – once engaged in radical ecological movements in Berlin, and later professor at the Free University – who began to see the potential of industry as a motor for social change towards a sustainable society (Huber, 1982).

On this background, Jänicke formulated a radical diagnosis of state failure as a parallel to the analysis of market failure. "State failure means supplying a country with public goods that are too highly priced and too low in quality" (Jänicke, 1990a: 1). According to Jänicke, the state's failure in providing public goods has three components. First, the state fails to take preventive action; it combats symptoms rather than causes (political failure). Second, the state fails to handle problems efficiently; public goods only come about as the result of excessive spending (economic failure). Third, the state's provision of public goods has not been very effective; the enormous growth in state budgets and administration stands in sharp contrast to the substandard quality of its activities (functional failure). Despite the apparent similarity to neo-liberal attacks on the state, Jänicke's perspective is certainly not right-wing: referring to the "economization of state activities", Jänicke sees state failures as a consequence of the economic forces generating market failures (1990a: 33-34). He ascribes much of the failure to the fact that bureaucratic-industrial complexes generate utility from the provision of public goods. In consequence, these interests are better off by pressing for increased spending and curative instead of preventive policies.

State failure has been most pronounced in the field of environmental policy (1990a: 41ff). The overriding problem is the political failure to address pollution problems at the source. The state has promulgated an enormous number of curative command-and-control regulations, aiming at end-of pipe treatment of pollution, while little has been done to promote cleaner technologies and other preventive methods. The problem with curative environmental policies is that they usually lead to problem displacements "from one medium or place to another – from water into the air or the ground, from densely populated areas to mountainous regions, or from industrial countries to the Third World ..." (1990a: 47). Drawing on evidence from a variety of countries, Jänicke comes to the following depressing conclusion concerning the environmental policies of the 1970s and 1980s:

> "After nearly 20 years the failure of environmental protection based on combating symptoms is plain to see. In some cases, such effects as have been recorded were achieved largely by redistributing the pollution, whilst others were to a significant degree not attributable to environmental protection measures but caused instead by the depression in smokestack industries, the high cost of energy and slowdown in the growth of the economy. In view of the environmental problems that were left unsolved and the new ones that arose, the costs were far too high. They cannot be justified even by saying that at least they led to some diminution in the high cost of damage to the environment" (1990a: 53).

Jänicke's devastating criticism is also a blow at the command-and-control solutions suggested by simplistic public goods theories, not least from the collective action perspective (cf. section 2.1). However, in the analysis of state failure, a positive outlook also appeared in the consideration of ecological modernization – an advanced stage of environmental policy, towards which some states have taken the first steps by stimulating the innovation of cleaner technologies (1990a: 52). The recognition that some countries have more capacity for entering this stage than others suggested the possibility of partial successes in environmental politics. Moreover, the evidence from cross-national comparative studies on structural and environmental change (Jänicke, 1990b; Jänicke and Weidner, 1995; Jänicke et al., 1996) indicated that some countries are almost invariably more effective than others in dealing with the environmental problems of their societies. The investigations of this problem led to the suggestion that the success of state policies on the environment crucially rests on the existence of certain institutional and structural conditions which shape the ecological modernization capacity of the state (Jänicke, 1990b; Jänicke, 1997). These capacity conditions at the macro-level are claimed to be of greater importance for environmental effectiveness than concrete policy instruments:

> "... successful environmental protection is brought about by a complex interaction of influences and not by a single, isolated factor, nor a favourite instrument, nor a single type of actor, nor a particular framework condition or institution. The literature on environmental policy and management is full of such proposals of a particular measure. The shortcomings of such a 'mono-factorial' view may be illustrated by the example of the time-consuming debate on environmental policy instruments (sometimes taking on the character of a debate on the 'instrument of the year'). Of course, we need assessments on the general advantages and weaknesses of our policy tools. But recent research makes it clear that this has been a simplistic debate using a mechanistic top-down model of policy and ignoring the complex interaction dynamics between quite different factors" (1997: 4).

But Jänicke did not stop at this point in the critique. In contrast to the bottom-up theorists, he actually set out to formulate a coherent theory on the specific institutional and structural factors conducive to the successfulness of environmental policies. According to this theory, a nation state's capacity for ecological modernization, and thereby its chance of successful environmental policies, depends above all on the following four structural and institutional factors (1990b: 222ff):

1. Economic leverage

Environmental policies are generally more successful in countries with strong economic performance. The correlation is not unambiguous. On the one hand, rich countries (measured in GDP per capita) are generally more motorized, electrified, chemicalized, etc. than poor countries; all factors resulting in heavy pollution loads. On the other hand, rich countries have at their disposal superior technological, material and institutional capabilities for pollution abatement. Moreover, the service sector in rich countries is growing compared to the industrial sector, and consumption patterns are affected by post-material values. On the whole, empirical studies confirm that pollution-retarding factors have the upper hand. However, a positive correlation between economic and environmental performance is conditioned by the availability of a standard technological solution (e.g. clean-up technology, substitutes, higher efficiency, or recycling) (1997: 14). In areas where a standard solution has not been developed (e.g. land use, soil contamination, road traffic), the opposite correlation will usually prevail, i.e. the higher GDP per capita, the more ecological deterioration.

2. Consensual capacity

Environmental policies are generally more successful in neo-corporatist countries characterized by a consensual policy style. In his cross-national empirical data, Jänicke finds a very strong correlation between the countries' degree of corporatism and their environmental progress on a number of selected indicators (1990b: 224). The six countries with the best environmental performance – Japan, the Netherlands, Luxembourg, Sweden, Switzerland and Austria – are all classified as strongly corporatist. The underlying causal mechanism is believed to be as follows: a consensual, cooperative style of interest-mediation, which is a hallmark of neo-corporatist countries, serves to weaken the effect of particularistic interests, thereby allowing for an earlier consideration of new broad-based ecological interests (1990b: 223). Moreover, compared to pluralist systems of interest-bargaining, corporatist systems of interest-mediation are seen as more capable of integrating overlapping policy concerns and taking a long-term perspective on the problems involved (1990b: 226, 228; 1997: 13). Both are important elements in shaping the strategic capacity of the state (see below).

3. Innovative capacity

Environmental policies are generally more successful in countries where the political and social processes are conducive to innovation (1990b: 224ff, 229). Progress towards ecological modernization requires the innovation of new environmental strategies and techniques. Innovative capacity depends, on the one hand, on social processes. Opportunity structures in the economic system may create fertile ground for industrial innovation of cleaner technologies. The institutionalisation of powerful information systems is equally important for innovative capacities. In particular, ecological modernization is stimulated by: openness of the scientific community and the media to new paradigms, problems and issues; effective communication channels for the diffusion of expert knowledge on cleaner technologies; and advanced environmental statistics and reporting systems (1990b; 1997). On the other hand, innovative capacities are also conditioned by processes at the state level, in particular the openness of political decision-making and the courts to new ecological ideas and interests. On this matter, Jänicke refers to the importance of "participative capacities" (1997: 12-13). The strong innovative capacities of the USA, for example, may explain its fairly good environmental performance in spite of weaknesses in consensual and strategic capacity (1990b: 225).

4. Strategic capacity

Environmental policies are generally more successful in countries with highly developed political-administrative competences for long-term planning and policy integration. Even if the aforementioned three capacities are present, it is not a sufficient guarantee for the success of environmental policies. Formulation and implementation of environmental policies require state-level institutionalisation in the form of administrative organizations, programmes, planning, resources and personnel. The institutionalisation gives rise to policy outputs that are more or less strategically appropriate in bringing about ecological modernization. According to Jänicke, strategic capacity generation is not so much a question of particular policy instruments, resources or personnel. Rather, the strategic capacity for ecological modernization is determined by the extent to which governmental institutions develop competences for, and give priority to, long-term planning and integration of various policy areas (1990b: 226; 1997: 13). Some countries are notably stronger than others regarding environmental planning and policy integration, e.g. Japan, Sweden, and the Netherlands, which all rank among the top four environmental performers in Jänicke's index. On the other hand, the superficial institutionalisation found in many South and East European countries creates difficulties for ecological modernization (1990b: 226).

The four structural-institutional factors should not be seen as isolated causal explanations. Taken together, they arguably constitute necessary and sufficient conditions for general success with respect to national environmental policies. But the factors are interrelated and to some extent mutually reinforcing. A systematic overview of cause-and-effect relations between the four clusters of variables is presented in graphic form in the original article (1990b: 228). Although strategic and innovative capacities are the most immediate causes of environmental policy output and outcome, the underlying causal drivers in the theory are economic leverage and consensual capacity. Economic leverage is central in defining the problem pressure, which again is the subject of strategic action and innovative efforts at the political-institutional level. Moreover, the economic structure affects the consensual capacity of a country through its impact on the organization of the main interest groups. Consensual capacity was originally seen as a socio-cultural factor by Jänicke, as it depends on social trust relations, but it is clear that it also incorporates a variety of institutional arrangements. A consensual, cooperative policy style is a very decisive condition for success: it promotes long-term orientation and policy integration (strategic capacities), and it allows for integration and early consideration of ecological interests in the political decision-making process (innovative capacity). Jänicke's essential point is therefore "that countries with relatively successful economic and labour market policies also tend to have relatively successful environmental policies" (Andersen, 1999).

In my view, the main problem with Jänicke's theory is its almost exclusive reliance on macro-level explanations. Most important in Jänicke's analysis are the factors that explain cross-national differences in environmental performance. But is it really true that some countries are invariably more successful than others with respect to their environmental performance? Jänicke's own classification is based on an index indicating the average progress in environmental quality of each country for a number of selected pollution problems over a selected time period. Some of his top performers are certainly not doing well in all areas. Japan, for example, which scores high, mainly due to its successful air pollution policies, still has considerable problems in managing water pollution, and the Netherlands – another top performer – is not doing well in agricultural pollution control or in current CO_2 pollution abatement. One might wonder what has become of Jänicke's original research agenda, according to which he was very sceptical of the nation-states' capacity for ecological modernization, and therefore looked for explanations of partial successes (Andersen, 1999). Partial successes refer to the few genuinely successful examples in specific sectors. Although Jänicke investigated partial successes via case study analyses, it had the effect of blurring rather than clarifying his argument, and it did not change the fact that his theoretical conclusions still centre on explanations of general successes.

The preoccupation with general instead of partial successes relates to a more substantial deficit in the Berlin school argumentation: how are the causes from structural institutional factors to environmental policy outcomes effectuated in concrete cases? Jänicke admits that the characteristics of policy instruments, actor interests, and the concrete situation play a role in effectuating the causes (1997), but he and the later adherents to the Berlin school are extremely silent on how these micro-level factors work out. The consensual capacity argument is a very good example of this. Jänicke simply claims that a neo-corporatist consensual policy style allows for earlier consideration of ecological interests, which in turn lead to ecological modernization. But are neo-corporatist systems of interest-mediation really as open to new interests as Jänicke suggests? The counterargument is that corporatism is a relatively closed system of interest mediation, where political power is monopolized by key associations in industry and labour. Jänicke points to the widespread consultation of environmental organizations in the corporatist Scandinavian countries, but forgets to say that they are often left out of the decisive negotiations between the social partners. Even more problems appear in Austria – by any reasonable standard, the most corporatist country in the world – where environmental organizations are to a large extent marginalized from governmental decision-making processes (Mol et al., 2000). Could we then fall back on the more plausible argument that corporatist institutions promote long-term orientation and policy integration? Perhaps[19], but again there is an under-specification of the causal mechanisms leading to successful environmental policy outcomes. Let me illustrate this point by examining some later contributions in the spirit of the Berlin school.

---

19. One counter-argument is that the close administrative coupling of implementation and regulation inherent in the corporatist system persuades regulators to compromise sustainability goals for short-term economic interests, creating a mismatch of technical and political issues at the implementation level which further erodes successful environmental policies (Hukkinen, 1995).

Based on Lundqvist's comparative study of clean air policies in Sweden and the USA (Lundqvist, 1980), and Martin Jänicke's theory, the idea that national corporatist institutions are conducive to successful environmental performance has been taken up by other authors (Crepaz, 1995; Scruggs, 1999). They have tested the thesis through statistical analyses of multi-country data, and have found strong positive correlations between variables measuring the countries' degree of corporatism and national performance on selected environmental issues. Crepaz claims that corporatism – defined as "a system of interest representation in which a small number of strategic actors (usually representatives of capital and labour), organized in peak associations, represents large parts of the population in an encompassing fashion" (Crepaz, 1995: 391-392) – facilitates efficient pollution abatement policies. This is supported by two theoretical arguments: First, corporatist networks between the social partners reduce transaction costs of establishing, overseeing and executing environmental policies. Second, corporatist organizations embody informal constraints that force the participants to behave in a cooperative manner, thereby making it easier to find solutions to the collective action problems associated with pollution reduction (1995: 398-399). Using a similar corporatism concept, with the addition of a policy concertation dimension, Scruggs also makes the general argument that corporatist institutions facilitate the overcoming of collective action dilemmas related to environmental public benefits. The main reasons are that the power of national peak associations facilitates the pursuit of national rather than particularistic interests, and secondly, that corporatist arrangements include effective schemes to compensate losers in the economic re-distribution resulting from environmental programmes (Scruggs, 1999: 5-6).[20]

Jänicke, Crepaz, and Scruggs deal almost exclusively with correlations at the macro-level and leave unspecified the macro-micro transitions that this causal relationship necessarily involves (from national corporatist institutions over the choice of regulatory instruments to the actors implementing cleaner technologies, and from local environmental effects to aggregate environmental performance). A central problem remains: how are individual polluters and groups of polluters encour-

20. Another important argument made by Scruggs is that corporatist institutions imply a history of producer-government trust in areas of industrial/-social policy, which increases acceptance of intrusive environmental regulation among the producers.

aged to undertake the concrete environmental improvements required to implement the general solutions agreed on by the social partners? The authors try to do away with the problem: "How particular producers or unions are 'sold' on the compromise may involve combinations of inducements, force or persuasion from the state or corporatist actors" (Scruggs, 1999: 30). Apart from the observation that good environmental performance presupposes the "shadow of strict regulation", i.e., the readiness of the policy-makers to regulate the activities of the producer groups if the pragmatic approach should fail (Scruggs, 1999: 5, 30; Crepaz, 1995: 398), policy instruments are nowhere included as an explanatory variable. The neglect of policy instruments raises a number of questions: does the tendency of corporatism to make room for better regulatory instruments explain its successfulness, or is it rather that corporatist institutions provide for better implementation of existing regulatory instruments? If so, are corporatist institutions really engines of implementation under all kinds of policy instruments? It is not true, for example, that the existence or non-existence of corporatist institutions matters very little to the incentive effects of green taxes. Questions like these need to be answered from a micro-/meso-level perspective that considers the institutional effects in their particular regulatory context.

While the Berlin school has contributed to a much better understanding of the social dynamics underlying environmental problems, and has increased our awareness of some of the institutional and structural factors facilitating effective environmental policies, it fails to account for the way in which environmental policy instruments affect the behaviour of polluters. Since the 1980s, policy instrument practices have expanded considerably compared to traditional command-and-control regulation. Although a limited number of systematic ex-post evaluations on the effectiveness of competing environmental policy instruments have been carried out so far, they nevertheless indicate that regulatory instruments are important to environmental effectiveness and are sometimes the decisive factor (e.g Knoepfel and Weidner, 1985; Andersen, 1994; Enevoldsen, 2000). The studies in which the superior effects of green taxes or tradable pollution permits compared to direct regulation have been demonstrated (for overviews see Hahn, 1989; OECD, 1997; Sprenger, 2000) call for a reconsideration of the theoretical basis underlying modern environmental governance. The evidence suggests that it is possible, albeit difficult, to design incentives that induce polluters to alter their behaviour in environmentally sound directions.

WHAT HAVE WE LEARNED FROM THE THEORIES? ENVIRONMENTAL GOVERNANCE AT THE BEGINNING OF THE 21ST CENTURY

As we have seen, rational theories in the fields of economics and political science have played an important role in the analysis of the political aspects of environmental problems, and have produced a number of normative prescriptions for good environmental governance. The primary strength of rational theories lies in their ability to put the incentives and capabilities of the polluters into a simple formula, allowing us to make universal generalisations on the behavioural effects of alternative pollution control instruments. The problem is that these predictions are not always very accurate. Sometimes this is due to unique circumstances or purely accidental factors. But there are certainly also more general reasons for the frequent failure of rational theory in predicting the effects of environmental governance. The real motives and capabilities of the individual actors differ considerably from ideal rationality assumptions; policy instruments do not pass undistorted through administrative apparatuses and target groups to the final policy outcome; and both of the aforementioned factors are to some extent shaped or affected by the institutional context in which they function. Insights on these matters are provided by institutionalist theories on environmental governance. On the other hand, purely institutionalist theories usually have difficulty in explaining behaviour at the micro-level; are individuals simply conforming to institutional roles, and even if they are, how do they resolve their many conflicting roles?

It appears from the critical discussion in the foregoing sections that there is no justification for claiming universal, or even general, superiority of the rational approach over the institutionalist approach, or vice versa. Each has improved our understanding of some aspects of environmental governance, while other aspects have been left unclarified. Milton Friedman defends the omissions in neo-classical rational choice theory with the fact that it has outperformed all other theories in making good general predictions (Friedman, 1995). But general predictive power is not the same as good prediction in concrete cases. It may be that rational choice theory offers the best prediction of very general and simple cases (e.g.: to what extent industry will pollute irrespective of the national and regulatory context). What is more interesting, however, is whether improved predictions can be made by alternative theories in specific cases (e.g: how well can we expect a voluntary environmental agreement between the Dutch chemical industry and the Ministry of Economic Affairs to work out, and will it be more effective than a similar arrangement in Denmark?). Environmental governance is precisely about such specific cases, and most of them cannot be adequately predicted without taking into account the contextual institutional factors. To have good predictions is therefore a matter of choosing the proper theoretical departure (rationalist or institutionalist) in the specific case and problem in question. But it does not need to be either/or; as I will argue in the following, it can be even more fruitful to combine the two approaches.

The descriptive analyses reviewed in this chapter have all contributed to the conceptual and theoretical basis for modern environmental governance. Simple rational choice theory, including collective action theory, played a central role in the formation and institutionalisation of modern environmental regulation in the late 1960s and 1970s (Weale, 1992). As time went by, the observed failures of command-and-control regulation sparked a variety of new political and theoretical developments. One consequence was an increased interest in economic instruments, especially Pigouvian taxes and tradable emission permits. The theoretical basis for this kind of governance was developed by the growing discipline of environmental economics. It is striking, however, that the tax-bounty solution proposed by Pigouvian externality theory – which dates back to the 1920s – was not practised until the late 1980s (apart from scattered incidents in water pollution policies). Another consequence was the development of the policy implementation theory, which offered more detailed explanations of what went wrong in the process of implementing traditional regulatory policy, and with recommendations on how to improve it (top-down or bottom-up). A third consequence was the emergence of an ecological modernization theory, focusing on the structural changes that would bring about sustainable development. Its implications for environmental governance were elaborated by the work of the Berlin school on the necessary structural and institutional capacities for successful environmental policies.

In this political and theoretical turmoil, and under the impetus of anti-authoritarian sentiments and new ideas of ecological modernisation, policy entrepreneurs and administrators began to develop entirely new solutions to complex problems of environmental regulation (including mediation, voluntary agreements, environmental management systems, eco-auditing, eco-labelling, etc.). These new instruments, which can be summarised under the term "joint environmental policy-making" (Mol et al., 2000), are positioned vis-à-vis other instruments in figure 1. In consequence, environmental governance of today is a patchwork of instrumental approaches with different theoretical foundations, some very specific, others very broad and ideological. This is not necessarily bad; a monolithic body of environmental governance based on a universal causal theory is neither realistic nor desirable in the complex world in which we live. Instead, we should try to clarify the unique areas in which individual approaches, or combinations hereof, establish useful causal relations between the mechanisms of environmental governance and environmental policy outcomes.

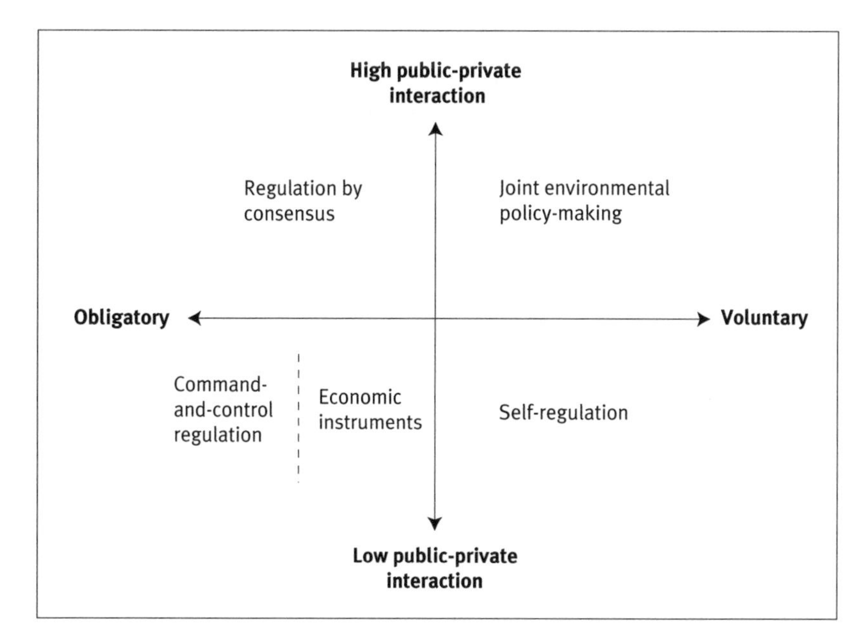

Figure 1. Types of environmental governance

Let us start with the collective action perspective on environmental governance. Paradoxically, this perspective is not very well positioned to understand the causal effect of the governmental intervention it recommends. The institutionalised and obligatory nature of the interventionist forms of regulation (see the two left quadrants of figure 1) leaves little scope for the non-cooperative game theory employed by collective action theory. However, with the emergence of new voluntary instruments in environmental governance (see the two right quadrants in figure 1), collective action theory deserves renewed attention. Specifically, it may shed light on the potential freeloader-rider problems of voluntary environmental agreements and similar instruments (Enevoldsen, 1999). On the other hand, as Ostrom already showed in the case of self-governing commons regimes, it clearly needs to be supplemented with an institutionalist perspective to account for the ways in which such problems can be overcome. The other rational approach is Pigouvian environmental economics. Its relevance is, of course, restricted to the economic instruments in the bottom-left quadrant, but transaction costs and other institutional factors need to be taken into account, even in analyses of the price mechanism (see e.g. Andersen and Sprenger, 2000).

Implementation problems must also be considered, which is where policy implementation theory enters the picture. It is particularly relevant in explaining the effects of traditional obligatory instruments (the left quadrants), as they usually involve a cumbersome implementation process. But to analyse the implementation of traditional policies without some ideas of the motives and interests of the actors seems futile, and some help could perhaps be found in the rational approach. Finally, there is the Berlin school of environmental governance, which in principle should apply to all forms of governance, as it concentrates on the contextual conditions. Nevertheless, evidence seems to indicate that countries with high ecological modernization capacities are more prone to make use of the sophisticated forms of governance based on high public-private interaction (the top quadrants) instead of command-and-control regulation or self-regulation (the bottom quadrants). The use of economic instruments is perhaps an exception to this rule. However, it does raise the question of whether it is structural-institutional background conditions or the choice of policy instruments that make a difference to effectiveness, and to resolve this question it will be necessary to integrate meso- and micro-level theories.

From the above discussion, it appears that each of the four theories has its own sphere of relevance, so none of them should be entirely rejected. At the same time, it also appears that each of the questions treated so far has a rational as well as an institutional element. Hence, it would perhaps be a good idea to pursue a closer integration between the theoretical approaches. One example of such closer integration is the rational institutionalist approach proposed by Elinor Ostrom (Ostrom, 1986; 1990; Ostrom et al., 1994). But it is clearly no more than a first step in this direction, and it is difficult to generalize the findings on the commons problem to other environmental policy areas. What seems to be needed is the development of a rational-institutionalist framework addressing the effectiveness of alternative forms of environmental governance, something I attempt to do in my forthcoming dissertation (Enevoldsen, forthc.).

Some degree of theoretical integration is also necessary in order to further confidence in the normative implications of the theories. At the present, collective action theory recommends command-and-control regulation or privatisation; environmental economists prescribe environmental taxes, bounties, and tradable emission permits; and the bottom-uppers and the Berlin school have a strong preference for negotiated rule-making and joint environmental policy-making (see, e.g. Weidner, 1998; Jänicke, 2000). Based on the past records of environmental policies and theoretical insights, we have learned that there is no one best instrument. However, it has also been obvious that some instruments have worked out better than others in specific cases. The problem, which integration of theories might be able to solve, is to specify the conditions under which it is preferable to choose one instrumental approach over another. However, we have learned from the Berlin school that good environmental governance is not only about choosing the right policy instrument and organising an effective implementation process. More efforts need to be devoted to capacity-building – especially the innovative, participative, integrative, and strategic capacities, which to some extent are changeable via the building of political institutions – so as to improve the background conditions for the practice of all types of environmental policy instruments (Jänicke, 1997).

REFERENCES

Andersen, M. Skou (1994). *Governance by green taxes: making pollution prevention pay*, Manchester: Manchester University Press.

Andersen, M. Skou (1999). Ecological Modernisation Capacity: Finding Patterns in the Mosaic of Case Studies. In: M. Joas and A-S. Hermanson (eds.), *The Nordic Environments: Comparing Political, Administrative and Policy Aspects*, Aldershot: Ashgate, pp. 15-46.

Andersen, M. Skou and R.-U. Sprenger (eds.) (2000). *Market-based Instruments for Environmental Management: Politics and Institutions.* Cheltenham: Edward Elgar.

Arrow, K. (1963) [second edition]. *Social Choice and Individual Values.* New Haven: Yale University Press.

Arrow, K. and F. H. Hahn (1971). *General Competitive Analysis.* San Francisco: Holden-Day.

Bardach, E. (1974). *The Implementation Game*. Cambridge Mass.: MIT Press.

Baumol, W. and W. Oates (1988). *The Theory of Environmental Policy*. Cambridge University Press

Bohm, P. and C.S Russell. Comparative analysis of alternative policy instruments. In: A.V. Kneese and J.L. Sweeney (eds.), *Handbook of Natural Resource and Energy Economics*, pp. 395-460. Amsterdam: Elsevier.

Borch, K. (1968). *The Economics of Uncertainty*. Princeton: Princeton University Press

Brennan, G. (1990). What Might Rationality Fail to Do? In: Cook and Levi (eds.), *The Limits of Rationality*, pp. 51-59. The University of Chicago Press.

Broome, J. (1990). Should a Rational Agent Maximise Expected Utility? In: Cook and Levi (eds.), *The Limits of Rationality*, pp. 132-145. The University of Chicago Press.

Buchanan, J. M. and G. Tullock (1962). *The Calculus of Consent*. Ann Arbor: University of Michigan Press.

Coase, R.H. (1960). The Problem of Social Cost. *Journal of Law and Economics*, No. 3, pp. 1-44.

Coase, R.H. (1988). *The Firm, the Market and the Law*. Chicago: The University of Chicago Press.

Cornes, R. and T. Sandler (1983). On Commons and Tragedies. *The American Economic Review*, Vol. 73, No. 4, pp. 787-792.

Crepaz, M.L (1995). Explaining national variations of air pollution levels: political institutions and their impact on environmental policy-making. *Environmental Politics* 4 (3), 391-414.

Dasgupta, P. (1990). The Environment as a Commodity. *Oxford Review of Economic Policy*, Vol. 6, No. 1, pp. 51-67.

Dawes, R.M. (1973). The Commons Dilemma game: An *N*-person Mixed-Motive Game with a Dominating Strategy for Defection. *ORI Research Bulletin*, Vol. 13, pp. 1-12.

Derthick, M. (1972). *New Towns in Town: Why a Federal Programme Failed*. Washington D.C.: Urban Institute.

Downing, P.B. and K. Hanf (eds.) (1982). *International Comparisons in Implementing Pollution Laws*. Dorchrecht: Kluwer-Nijhoff Publishing.

Downing, P.B. and L.J. White (1986). Innovation in Pollution Control. *Journal of Environmental Economics and Management*, Vol. 13, pp. 18-29.

Downs, A. (1957). *An Economic Theory of Democracy*. New York: Harper & Row.

Dunsire, A. (1978). *Implementation in a Bureaucracy*. Oxford: Martin Robertson.

Elmore, R. (1979). Backward Mapping: Implementation Research and Policy Decisions. *Political Science Quarterly*, Vol. 94, No. 4, pp. 601-616.

Elster, J. (1979). *Ulysses and the Sirens: Studies in rationality and irrationality*. Cambridge University Press.

Enevoldsen, M. (1999). *The effectiveness of voluntary agreements: a theoretical analysis of free rider problems*. The Department of Political Science, Aarhus University.

Enevoldsen, M. (2000). Industrial Energy Efficiency. In: A.P. Mol, V. Lauber and J.D. Liefferink (eds.), *op.cit.*, pp. 62-103.

Enevoldsen, M. (forthc.). *Environmental agreements and taxes: a new institutionalist approach to effective* CO_2 *regulation*. Ph.D.-dissertation in progress, The Department of Political Science, Aarhus University.

Friedman, M. (1953). *Essays in Positive Economics*. Chicago: University of Chicago Press.

Gunn, L. (1978). Why is implementation so difficult? *Management Services in Government*, Vol. 33, pp. 169-176.

Hahn, R.W. (1989). Economic prescriptions for environmental problems: how the patient followed the doctor's advice. *Journal of Economic Perspectives*, Vol. 3, No. 3, pp. 95-114.

Hanf, K. (1982). Regulatory Structures: Enforcement as Implementation. *European Journal of Political Research*, Vol. 10, pp. 159-172.

Hanf, K. and F. Scharpf (eds.) (1978). *Interorganizational Policy Making: Limits to Coordination and Central Control*. London: Sage Publ.

Hammond, T.H. (1996). Formal theory and the Institutions of Governance. *Governance: An International Journal of Policy and Administration*, Vol. 9, No. 2, April, pp. 107-185.

Hanley, N. and H. Folmer (eds.) (1998). *Game Theory and the Environment*. Cheltenham: Edward Elgar.

Hardin, G (1968). The Tragedy of the Commons. *Science*, Vol. 162, pp. 1243-1248.

Hardin, R. (1971). Collective Action as an Agreeable *N*-Prisoner's Dilemma. Behavioral Science, Vol. 16, pp. 472-481.

Hardin, R. (1982). *Collective Action*. Baltimore: John Hopkins University Press.

Heap, S. Hargreaves, M. Hollis, B. Lyons, R. Sugden, A. Weale (eds.) (1992). *The Theory of choice: A Critical Guide*. Oxford: Blackwell Publishers.

Hjern B., K. Hanf and D. Porter (1978). Local Networks of Manpower Training in the Federal Republic of Germany and Sweden. In: K. Hanf and F. Scharpf (eds.), *op. cit.*, pp. 303-344.

Hjern, B. and C. Hull (1981). Implementation Research as Empirical Constitutionalism. *European Journal of Political Research*, Vol. 10, pp. 105-115.

Hood, C.C. (1976). *The Limits of Administration*. London: John Wiley.

Huber, J. (1982). *Die verlorene Unschuld der Ökologie: Neue Technologien und Superindustrielle Entwicklung*. Frankfurt am Main: Fischer Verlag.

Hukkinen, J. (1995). Long-term environmental policy under corporatist institutions. *European Environment*, Vol. 5, No.4, pp. 98-105.

Jänicke, M. (1990a). *State Failure: The Impotence of Politics in Industrial Society*. Cambridge: Polity Press.

Jänicke, M. (1990b). Erfolgsbedigungen von Umweltpolitik im Internationalen Vergleich. *Zeitschrift für Umweltpolitik*, No 3., pp.. 213-232 [English version: Jänicke, M. (1992). Conditions for Environmental Policy Success: An International Comparison. *The Environmentalist*, Vol. 12, No. 1, pp. 47-58].

Jänicke, M. (1997). The Political System's Capacity for Environmental Policy. In: M. Jänicke and H. Weidner (eds.) (1997), *National Environmental Policies: A Comparative Study of Capacity-Building*, pp. 1-24. Berlin: Springer.

Jänicke, M. (2000). Environmental innovations from the standpoint of policy analysis: from an instrumental to a strategic approach in environmental policy. In M. Skou Andersen and R.U. Sprenger (eds.), *op.cit*, pp. 49-66.

Jänicke, M. and H. Weidner (eds.) (1995). *Successful Environmental Policy: A Critical Evaluation of 24 Cases*. Berlin: Edition Sigma.

Jänicke, M., H. Mönch and M. Binder (1996). Umweltindikatorenprofile im Industrieländervergleich: Wohlstandsniveau und Problemstruktur. In: M. Jänicke (ed.), *Umweltpolitik der Industrieländer: Entwicklung – Bilaz – Erfolgsbedingungen*, pp. 113-131. Berlin: Edition Sigma.

Katzenstein, P. (1984). *Corporatism and Change: Austra, Switzerland and the Politics of Industry*, Ithaca.

Kemp, R. (1997). *Environmental Policy and Technical Change: A Compar-*

ison of the Technological Impact of Policy Instruments. Cheltenham: Edward Elgar.

Knoepfel, P. and H. Weidner, (1985). *Luftreinhaltepolitik im internationalen Vergleich*, Band 1-6. Berlin: Edition Sigma.

Knudsen, C (1997) [second edition]. *Økonomisk metodologi; bind 2: Virksomhedsteori og industriøkonomi*. København: Jurist og Økonomforbundets forlag.

Kooiman, J. (ed.) (1993). *Modern Governance: New Government Society-Interactions*. London: Sage Publ.

Lauber, V. (2000). The Political and Institutional Setting. In A.P. Mol, V. Lauber and J.D. Liefferink (eds.), *op.cit.*, pp. 32-61.

Lewin, L. (1991). *Self-Interest and Public Interest in Western Politics*. Oxford: Oxford University Press.

Libecap, G.D. (1989). *Contracting for Property Rights*. Cambridge: Cambridge University Press.

Lipnowski, I. and S. Maital (1993). Voluntary provision of a pure public goods as the game of chicken. *Journal of Public Economics*, Vol. 20, pp. 381-386.

Lipsky, M. (1979). *Street Level Bureaucracy*. New York: Russell Sage Foundation.

Lundqvist, L. (1980). *The Hare and the Tortoise: Clean Air Policies in the United States and Sweden*. Ann Arbor: University of Michigan Press.

Mansbridge, J. (1995). Rational Choice Gains by Losing. *Political Psychology*, Vol. 16, No.1, pp. 137-155.

March, J.G. and J.P. Olsen (1984). The New Institutionalism: Organizational Factors in Political Life. *American Political Science Review*, Vol. 78, pp. 734-749.

Margolis, H. (1982). *Selfishness, Altruism and Rationality*. The University of Chicago Press.

Mayntz, R. et al. (1978). *Vollzugsprobleme der Umweltpolitik*, Wiesbaden: Kohlhammer.

Mäler, K-G (1989). The Acid Rain Game. In: H. Folmer and E. van Ierland (eds.), *Valuation Methods and Policy Making in Environmental Economics*. Amsterdam: Elsevier.

Milliman, S.R. and R. Prince (1989). Firm Incentives to Promote Technological Change in Pollution Control. *Journal of Environmental Economics and Management*, Vol. 17, pp. 247-265.

Mol, A.P., V. Lauber and J.D. Liefferink (eds.) (2000). *The Voluntary Approach to Environmental Policy: Joint Environmental Approach to Environmental Policy-Making in Europe*. Oxford: Oxford University Press.

Morgenstern, O. (1935) [English translation published 1976]. Perfect foresight and economic equilibrium. In: A. Schotter (ed.), *Selected Economic Writings of Oskar Morgenstern*, pp. 169-183. New York: New York University Press.

Mueller, D. C. (1989). *Public Choice II.* Cambridge University Press.

North, D. (1990). *Institutions, institutional change and economic performance.* Cambridge University Press.

OECD (1997). *Evaluating economic instruments for environmental policy-making.* Paris: OECD.

Olson, M. (1965). *The Logic of Collective Action.* Cambridge, Mass: Harvard University Press.

Ostrom, E. (1990). *Governing the Commons: The Evolution of Institutions for Collective Action.* New York: Cambridge University Press.

Ostrom, E., R. Gardner and J. Walker (1994). *Rules, Games and Common-Pool Resources.* Ann Arbor: The University of Michigan Press.

Pearce, D.W. and R.K Turner (1990). *Economics of Natural Resources and the Environment.* London: Harvester Wheatsheaf.

Pigou, A.C. (1920) [fourth edition, 1946]. *The Economics of Welfare.* London: MacMillan and Co.

Pressman, J.L. and A. Wildawsky (1973). *Implementation: How Great Expectations in Washington Are Dashed in Oakland; Or, Why It's Amazing that Federal Programs Work at All, This Being a Saga of the Economic Development Administration as Told by Two Sympathetic Observers Who Seek to Build Morals on a Foundation of Ruined Hopes.* Berkeley: University of California Press.

Rasmussen, (1994) [second edition]. *Games and Information: An Introduction to Game Theory.* Cambridge Mass.: Blackwell Publishers.

Riker, W. H. and P. C. Ordeshook (1973). *An Introduction to Positive Political Theory.* New Jersey: Prentice-Hall.

Sabatier, P. (1986). Top-Down and Bottom-Up Approaches to Implementation Research: a Critical Analysis and Suggested Synthesis. *Journal of Public Policy*, Vol. 6, No. 1, pp. 21-48.

Sabatier, P. and D. Mazmanian (1979). The Conditions of Effective Implementation: A Guide to Accomplishing Policy Objectives. *Policy Analysis*, Vol. 5, pp. 481-504.

Sabatier, P. and D. Mazmanian (1983) [1989]. *Implementation and Public Policy: with a New Postscript.* New York: University Press of America.

Sabatier, P. and H. Jenkins-Smith (1993). *Policy Change and Learning: An Advocacy Coalition Approach.* Boulder: Westview Press.

Sabatier, P. and H. Jenkins-Smith (1998). The Advocacy Coalition Framework: An Assessment. In: P. Sabatier (ed.), *Theories of the Policy Process*, pp. 117-168. Boulder: Westview Press.

Sandler, T. (1992). *Collective Action: Theory and Applications*. London: Harvester Wheatsheaf.

Schelling, T. C. (1978). *Micromotives and Macrobehavior*. New York: W.W Norton & Company.

Schelling, T. (ed.) (1983). *Incentives for Environmental Protection*. Cambridge Mass: MIT Press.

Schotter, A. (1981). *The Economic Theory of Social Institutions*. Cambridge: Cambridge University Press.

Scruggs, L.A. (1999). Institutions and Environmental Performance in Seventeen Western Democracies. *British Journal of Political Science, Vol. 29, pp. 1-31.*

Sen, A. (1982). Rational Fools. In: A. Sen, *Choice, Welfare and Measurement*. Oxford: Blackwell Publishers.

Simon, H. (1957). *Administrative Behaviour: A Study of Decision Making Processes in Administrative Organization*. Second edition. New York: Macmillan.

Sprenger, R.-U. (2000). Market-based instruments in environmental policies: the lessons of experience. In: M. Skou Andersen and R.U. Sprenger (eds.), *op.cit.*, pp. 3-27.

Stigler, G. J. and G. S. Becker (1977). De gustibus non est disputandum. *American Economic Review*, Vol. 67, No. 1, pp. 76-90.

Taylor, M. (1987). *The Possibility of cooperation: studies in rationality and social change*. Cambridge University Press.

Taylor M. and H. Ward (1982). Chickens, whales, and lumpy goods: alternative models of public-goods provision. *Political Studies*, Vol. 30, pp. 350-370.

Tietenberg, T.H. (1990). Economic instruments for environmental regulation. *Oxford Review of Economic Policy*, Vol. 6, No. 1, pp. 17-33.

von Neumann, J. and O. Morgenstern [1944], (1980). *Theory of Games and Economic Behavior*. Princeton: Princeton University Press, 3. Edition (originally published 1953, 1. Edition originally published 1944).

Weale, A. (1992). *The new politics of pollution*. Manchester: Manchester University Press.

Weidner, H. (ed.) (1998). *Alternative dispute resolution in environmental conflicts: experiences in 12 countries*. Berlin: Edition Sigma.

CHAPTER 4

The Ethical Rationale of the Concept of Sustainability, and Rationality Conflicts in Environmental Law

Ulli Zeitler and Ellen Margrethe Basse

SUSTAINABILITY AND ENVIRONMENTAL LAW: ASSUMPTIONS ON RATIONALITY

CONSIDERATIONS OF sustainability have been well known for over 20 years in connection with the legal and political administration of a common concern for the preservation of the environment and natural resources. The concept of sustainability is primarily an expression of a development ambition adapted to prevalent environmental conditions, i.e. of balancing ecological, economic and social concerns. The "principle of integration", which states "that development plans should be compatible with a sound ecology and that adequate environmental conditions can best be ensured by the promotion of development" (Sands, 1995 (b), p.61), is probably the core idea of sustainability in its various political implementations.

Demands for sustainability are included in the great majority of national and international environmental laws and declarations passed since the early 90s, and are clarified to various degrees in environmental principles, guidelines and agendas. Since central environmental concerns have a trans-boundary character, the main focus has been on international environmental law and policies. However, even on a global scale, demands for sustainability must come to terms with specific cultural characteristics, differing national legislation, current political relationships, and local production and consumption patterns. Otherwise there is a fatal lack of compliance and effectiveness. Inappropriateness to local and national conditions is a permanent source of conflict.

Moreover, even if some environmental problems have global significance, concrete action must normally be taken at a local level.

It is thus imperative to analyse the concept of sustainability in order to bring to light the implicit local and cultural interpretations, potential lack of accord and social discrepancies, which combine to form a major obstacle to the promotion of coherent international and national environmental regulation. Different rationalities may ultimately undermine the global project of promoting sustainable practices unless these differences are given due respect in the formulation and implementation of sustainable strategies.

The positive rhetoric of sustainability has been relatively effective in hiding its basic, partly contradictory presumptions (based on different ethical values, political goals, ecological theories, and economic interests), and the potential conflicts between the different aspects of environmental law and other parts of the legal framework. The indetermination of largely uniform texts on international environmental agreements is politically convenient, but probably also inefficient. The important question is to what extent uniformity is necessary or even desirable, and, consequently, how should the need for a unanimous interpretation of sustainability be described. If unanimity is purchased at the price of lack of implementation, different approaches might be preferable. Perhaps interpretation should be left open, thereby ensuring compliance and adequate implementation by various local and national cultures, even if this prevents materially specific international agreements.

The current international consensus on sustainability is based on the interpretation embodied in the United Nations World Commission on Environment and Development (the Brundtland) Report and several conventions, declarations, etc. The documents may rely on alternative interpretations and suffer from internal contradictions. Yet, they share WCED's vague idea of an ecologically sound economic development, based on intra- and intergenerational justice. The emphasis on satisfaction of needs, distributive justice and a sound relationship to the environment underlines a heavy reliance on ethical assumptions as an integral part of economic resource management and legislation.

The dominant interpretation of the underlying ethics is clearly anthropocentric (as opposed to ecocentric), insofar as the WCED had the interests of human beings, and only these, in mind. Legal documents treating sustainability claims, however, are not very clear on this point; or, at least, they want to keep the specific interpretation open (Hohmann, 1994, p. 3; Calliess, 1997). Although from a pragmatic *short term point of view*, specific interpretation may seem irrelevant with respect to the solution to pressing environmental problems, it remains to be investigated how anthropocentric and ecocentric interpretations may affect society and nature in a *long term* perspective, and in relation to natural conditions which are not yet regarded as critical challenges.

This perspective is reflected in the idea of inter-generational justice, which defines the content of sustainability. However, there are numerous practical and theoretical problems to be dealt with here. In the first place, there is neither a clear answer to the question of what is a proper time scale, nor to how one can identify people who do not and may never exist. Secondly, handling the needs of collectives such as present and future generations may be difficult to accomplish within a tradition of individual responsibility. Thirdly, the current focus on law and order requirements, being mainly concerned with the protection of those who are directly affected, causes serious difficulties for the implementation of legal and administrative policies on sustainability. Therefore, to ameliorate the discussion of sustainability, we must clarify relationships of responsibility in their legal and moral context.

This article presents an analysis of the legal and ethical rationalities which substantiate different interpretations of sustainability in legal and political documents and are responsible for frictions and dilemmas in environmental management. It will point out how divergencies in interpretation affect the implementation of cross-national environmental regulation.

A proper forum for the analysis will be the highlighting of various rationality conflicts and problems of cross-cultural communication, conceptual issues, and institutional conditions, as found in Europe, New Zealand and Japan, respectively. After a short delineation of different legal and ethical rationalities, the integration of sustainability claims in the legal frameworks of the EU, New Zealand and Japan is described, followed by an analysis of their ethical rationale, concentrated on the features of anthropocentrism, intrinsic values, and pragmatism. In the concluding part of the article, the possibility of a global consensus, based on cross-cultural communication, will be discussed.

Legal notions and rationalities

A legal system is an operating set of legal institutions, procedures, and material rules (Merryman et al., 1994, p.3). Nations which are substantially different in their orientation are likely to have divergent legal systems, based on different rationalities. The legal tradition is not only a set of rules. Rather it is a set of deeply rooted, historically conditioned attitudes to rationality, the role of law in society and the civil constitution, the language of the people, the proper organisation of a legal system and traditions on the way law should be applied (ibid.). Most important in this context is the understanding of the rights and duties of individuals and their conception of nature, as this is essential for effective environmental regulation.

In general, an action or institution is said to be "rational" if it satisfies the (overt or implicit) expectations of the social community. In this abstract sense, the definition fits all the cases discussed in the following. (For an overview over types of legal rationalities, see Bengoetxea, 1993, pp. 175-180).

Europe

The civil, common, socialistic legal systems express ideas and embody institutions which have been formed in the European historical and cultural context (Merrymann et. al, 1994, p. 7). European legal concepts are typically framed in capitalist democracies. This legal and jurisprudential tradition has as its central task the solving of specific conflicts on the basis of general rules and a formal principle of justice according to which we should treat like situations equally. In such legal systems there is growing uniformity on the definition of individual rights, making *protection of the private property interest* the most important concern. Protection is guaranteed in constitution as well as by statutes and case law. Accordingly, nature can be the object of legal regulation in terms of rights and obligations to own and use natural resources.

A proper, rational attitude towards the environment is an attitude formed by law-abiding adult citizens who have an interest in a particular piece of nature on which they can make a legal claim. The nature of this claim is expressed in different rationalities in the history of legal reasoning. In the tradition of *legal positivism*, a sharp distinction is made between law – including law based on a common rationality – and the morality of society. The latter becomes relevant only as the context within which, law, as an expression of the norm-constituting intentions of the legislator, is materialised. In contrast, the tradition of *natural law* sums up a variety of legal philosophies which make the legitimacy of legal claims dependent on extra-legal sources. One presupposition has been the claim that human beings have certain common natural characteristics which lead to similarities in social structures, including law and legal systems (Merryman et al. 1994, p. 18). This gives morality a central importance in relation to judicial norms, ideals, principles and values. For modern European environmental law and its global aspirations, a natural law approach based on the moral foundation of human rights and obligations is most convenient.

As a constitutive part of modern Western culture, jurisprudence and legal institutions share a dependence on abstract principles and concepts. Among such ideas a strong belief in the universal rights of human beings (independent of their particular natural context) and in the reality of nature as a meaningful entity transversing specific manifestations is stressed. According to the European mind, we are all assumed to share these beliefs and their legal expressions in spite of cultural diversities.

New Zealand

The population of New Zealand is faced with two different rationalities of law, relating to the European tradition and the Maori culture, respectively. The first historical attempt to reconcile these traditions by the "Treaty of Waitangi" is still a highly contested issue. With the Resource Management Act (RMA) of 1991, the disputes have been revived. Apparent consensus on a specific legal document is based on divergent interpretations of central concepts, and draws a veil over basically incompatible values (Harris, 1993, p. 73). The lack of a common value-scale is already indicated by the discrepancy between the European ideal of value-freeness inherent in the claim of global consensus, i.e. "a rhetoric of impartiality and equality" trying "to deny the reality of its cultural bias and its political servitude" (Jackson, 1992, p. 6), and the Maori rationality of law ("te maramatanga o nga tikanga") based on kinship relations, i.e. a hierarchical power relationship.

Moreover, the foundation of Western legal practice in individual property rights is in opposition to Maori custom. According to the Maori, it is impossible to possess land in any familiar legal sense. The Maori call themselves "tangata whenua", "people of the land" or "natives", i.e. in relation to the soil, they are "born of it, their generations buried in it, attached to it by indissoluble spiritual ties" (Sharp, 1990, p. 8). Their land is their sphere of "mana" (authority), which is also their sphere of accountability. Something happening within one's sphere of accountability, whether causally responsible or not, is nevertheless something for which this person (or the collective) is blamed. Consequently, the sphere of accountability, or "mana", cannot be legally determined by a contract or by a statute on rights and duties as in European law, but is historically, genealogically ascertained ("whaka-papa").

The imposition of an environmental law on Maori culture may be particularly problematic, due to its distortive effect on an already well-functioning environmental management with solid historical roots. Maori environmental practice is based on an idea of guardianship ("Kaitiakitanga"), which is different from a corresponding idea in European legal thought, where it is related to individual property rights. As a *legal* concept and principle, "Kaitiakitanga" needs a legally established principle of collective accountability and the acknowledgement of the legal status of natural phenomena in order to be operable. Rules securing legal status for these values, however, would not only be markedly different from what is known in European legal systems, but would threaten a less legalistic, well-functioning social practice.

Japan

Japan was the only non-Western, independent nation which – based on European models – undertook a comprehensive reform of its legal system during the 19th century (Merrymann et. al., 1994, p. 7). Today, Japan's legal system remains solidly embedded in this tradition. Nevertheless, the adaptation of European legal concepts has not led to the disregard of national traditions and cultural values (Chiba, 1995, p. 76). In spite of apparent similarities to European schemata, the Japanese perception and conceptualisation of problems may diverge significantly from them.

One distinctive and still important feature of Japanese culture is the lack of a solid dualistic tradition according to which referents (signifié) are automatically objectified. Within a non-dualistic culture, what are referred to are particular events *as lived,* not objectified, i.e. as something of which the language user forms a constitutive, transforming part. This also implies that man is basically not an individual subject, but has a trans-individual, historical and genetically constituted identity, which comes before the individual and out of which everyone is born. As a consequence, any Far-Eastern culture has to struggle with the introduction of European ideas of individual human rights, which in many cases are in opposition to concepts based on trans-personal identity and common interest (Yamagami & Suketake, 1994, p. 19).

Authentic moral virtues and legal principles – which in Western thinking are rooted in Divine revelation and individual, conceptual rationality – in Japanese culture are located in the specific "between man and man" (Kimura 1995, pp. 14-17). The focus on the interdependence of life, on the "between", in contrast to individuals, fosters the concept of collective guilt and responsibility, which distinguishes this form of rationality from Western ideas of individual responsibility and causally determined moral complaints. The concept of collective responsibility implies both that "actor" and "victim" are affected in a way which not only excludes one-sided responsibilities, but also transcends causal relationships. "Between one's guilt and the other's misfortune there is no room for causal relations. Both are indistinguishable dimensions of the same event." (Kimura 1995, p. 51, author's translation)

A person's shame and moral and legal responsibility are directed to the basis of his or her existence, the "between man and man", not to an identifiable individual. Thus, the convicted criminal or transgressor is not just an agent who must be held responsible, but a representative of a socially unacceptable context of action.

The Japanese legal system in its present European form is a pragmatic institution, adopted as part of the rapid Westernisation following the Meiji revolution in 1868. Apart from its agreement with pragmatism, which is a natural part of the Shinto-Buddhist heritage, the import of Western institutions has no natural basis in Japanese culture. Neither has the recent preoccupation with environmental law.

Lacking abstract notions, the Japanese language had originally no word for the Western concept of "nature", but a great variety of expressions for specific natural phenomena (Ono 1966, p. 94). The concept of "environment", although well established in contemporary Japanese society, also lacks a cultural foundation. It has a highly technical meaning in current literature, reflecting the usage predominant in Western societies. As a part of international society, Japan intends to fulfil its obligations in relation to environmental protection, and therefore to make rules for the environment. Yet, discrepancies between abstract, scientific demands (abstract rationality) and specific local and national practices (pragmatic rationality) in Japan must not be overlooked.

Types of ethical rationality

Although legal rationalities are deeply influenced by moral attitudes, the legal focus on *external* regulation and *enforceable* sanctions gives them certain distinctive features. Ethics, less concerned with pragmatic questions of social co-ordination of actions than with questions of the rightness of attitudes and actions, has its own standards, although they are dependent on institutional, historical and biological circumstances.

There are many ways of classifying ethical rationalities. The main types to be focused on in the following are so-called first-order ethical rationalities, i.e. basic ways for human beings (as moral agents) to structure experiences and to form life expectations. These first-order rationalities are fundamental to second-order distinctions between different ethical theories like utilitarianism, Kantianism, virtue ethics, or phenomenological ethics. For the purpose of a trans-cultural, comparative study, however, the rationalities of Western ethics do not provide an adequate framework for analysis. A meta- or trans-ethical level of analysis is required, particularly as some of the societies included in this analysis do not have a tradition for philosophical and ethical investigations.

The two first-order distinctions to be made are those of *centrism* and *reflectivity,* respectively. The perception of natural conditions and the interpretation of sustainability demands and environmental law are related to the way humans centre this information, and to their level of reflectivity, i.e. the involvement of conceptual or bodily faculties.

First, our actions may be divided according to their general perspective or outlook into (1a) *anthropocentric thought*, which makes human interests the focus and criterion of all activity, and (1b) *ecocentric thought*, which makes human interests a sub-category of "nature's interests", assuming the possibility of "intrinsic" or "inherent" values of non-human natural phenomena (animals, plants, ecosystems etc.). Although anthropocentrism, in a trivial sense, is unavoidable – insofar as any structuring by humans as moral agents must arise from human capabilities – in an important and morally relevant sense, the inclination to recognise only human interests as worthy of moral concern (genuine anthropocentrism) has no convincing rational foundation. In this study, the anthropocentrism of Western societies – e.g. as European life world and institutions – is contrasted with the alleged ecocentrism of Maori people and the transcendence of both anthropocentrism and ecocentrism into centreless thinking as an expression of certain basic features of Japanese culture.

Secondly, our actions may be distinguished according to their origin in reflectivity or bodily experience: (2a) rationality in its proper sense *as reflective activity,* which generally follows rules of deductive and inductive logic, including abstract, and certain forms of pragmatic, rationality; (2b) *non-reflective*, non-verbal and non-logical ways of encountering, which include practical involvement, bodily experience and intuition.

Although they may not satisfy certain scientific criteria for logical reflection, these non-rational forms of experiencing and acting may be quite "reasonable" in a practical perspective. At least they have *sense* and *meaning,* which is at least a minimum, and probably sufficient, condition for rationality.

THE INTEGRATION OF SUSTAINABILITY IN LAW

The concept of sustainability has found a common source of reference across different cultures and rationalities in the Brundtland Report. Although an unambiguous definition is lacking, the report established a connection between new economic growth and development possibilities and genuine environmental improvements. Thus the policy and legal implementation of sustainability is not a sectoral issue but a general political and socio-economic task.

> "The international law of sustainable development is ... broader than international environmental law; apart from environmental issues, it includes the social and economic dimensions of development, the participatory role of major groups, and financial and other means of implementation." (Sands, 1995 (a), p. 14)

Legal requirements for sustainable solutions in the sense of a "good environment" include citizens' access to information on issues of importance to their health and welfare, but probably also legal representation of the interests of non-rational beings, including animals, plants and ecosystems. Such legal demands are today partly expressed and formalised in rules on *E*nvironmental *I*mpact *A*ssessments (EIA), *S*trategic *E*nvironmental *A*ssessments (SEA), and on the obligation of authorities and other responsible institutions to publish information on the environmental situation. The purpose of these instruments is to secure an integrated preliminary assessment at policy, programme, plan, and project levels. International standards for such assessments have been laid down, such as that proclaimed in the 10th principle of the Rio Declaration:

> PRINCIPLE 10. "Environmental issues are best handled with the participation of all concerned citizens, at the relevant level. At the nation-

> al level, each individual shall have appropriate access to information concerning the environment that is held by public authorities, including information on hazardous materials and activities in their communities, and the opportunity to participate in decision-making processes. States shall facilitate and encourage public awareness and participation by making information widely available. Effective access to juridical and administrative proceeding, including redress and remedy shall be provided."

Administratively, the sustainability concept implies better co-ordination, increased co-operation, and integration of environmental principles, considerations, and management in all private and public decision-making systems. When "capacity-building" is used as a consensus-creating tool and is expected to produce effective results in the implementation of international demands for sustainability, it also implies a wide participatory approach, which includes governments and their partners, such as industry, NGOs, the public, etc.

Despite verbal commitments to holistic solutions in discussions of sustainability, sectoral treatments are still the most common reactions. The concept of environment, and a natural science approach to environmental issues, invite sectoral thinking and form major obstacles to fruitful trans-boundary discourse.

A central concept which unites socio-economic and environmental concerns is the idea of *equity* or distributive justice. However, even considerations of equity do not share a common global understanding. One must distinguish between a West-orientated approach, based on individual property rights, and the idea of a trans-individual, historical, genetically constituted equity. To what extent non-human entities may form part of proper equity considerations and intra- and intergenerational distributive justice, remains, however, disputed.

National and international laws reflect quite different problems and discussions. Basically, Europeans must struggle with the dilemmas and problems of an anthropocentric culture and the permanence of individual property rights. New Zealand's legislation, forced to reconcile two incompatible cultures, highlights the conflict between two quite different ethical rationalities: anthropocentrism and ecocentrism. Conceptualisations in Japanese law, furthermore, while resembling that of Europe on the surface, nevertheless lie somewhere beyond the distinction between anthropocentrism and ecocentrism, and draw attention to a radically different legal and political approach.

Contrary to the traditional concept of law and order, the concept of sustainability is not primarily based on a need for protection of individual rights – e.g., against a strong public authority – but rather on a need for protection of the human species, the availability of the resources for human well-being, and the rate of consumption of these resources. While different interpretations and legal traditions may prevent unanimous global implementation, they may still, in different ways, promote sustainable practices.

Sustainable development and the market in the European Union

This article uses the European Community to illustrate the Western tradition. Therefore, it must be stressed that there are several differences between US and European environmental law (Rehbinder and Stewart, 1988). The former, which is based on the tradition of common law, uses legal forms of collective responsibility which are largely unknown in European law. The latter, and the laws of the Member States of the European Community (especially Denmark) better illustrate the importance and consequences of a legal tradition based on an individualistic approach.

At the Community level, the principle of sustainability has been demonstrated in particular with the Treaty on the European Union in November 1993. With Article 2, the concept of sustainability was introduced in the context of the objectives of a common market,

> "... to promote throughout the Community a harmonious and balanced development of economic activities, sustainable and non-inflationary growth respecting the environment, a high degree of convergence of economic performance, a high level of employment and of social protection, the raising of the standard of living and quality of life, and economic and social cohesion and solidarity among Member States."

Sustainable development is replaced by the principle of "sustainable and non-inflationary growth". The emphasis on "solidarity among Member States" and the objective of "raising" the "standard of living" in one of the World's richest areas, however, indicate a substantial distance from the Brundtland Report and the Rio Declaration, i.e., from the idea of global sustainable development and the implicit demands for equity.

Shortly after the adoption of the Maastricht Treaty, the 5th Action Programme "Towards Sustainability" was passed. The programme uses the loaded concept "sustainability" as a basis for laying down the normative and institutional framework as well as the means of realising the European objectives of a single market and of securing continued economic growth and social development in the Community. Its aim is to properly manage economic growth and social improvements without detriment to the environment. It is emphasised that:

> "the wanted balance between man's activities and development and the environmental protection demands a fair distribution of responsibility concisely defined as to consumption and behaviour towards the environment and the natural resources ..." (Official Journal, C 138, 1993, p. 11 litra 4).

The demand for sustainability expresses a need for alternative thinking in considering the concept of "responsibility". A more broadly based, active participation of all the economic and social actors is regarded as a basic requirement for achieving sustainability.

The Treaty was again amended by the Treaty of Amsterdam in 1997. Article 2 describes the objectives of the Community. One of these is "to promote economic and social progress and to achieve balanced and sustainable development", without explaining what balanced and sustainable development is (Krämer, 2000).

When utilising the concept of sustainable development or sustainable growth, *the balance of interests* depends on the interplay between formal considerations of equality, individualistic jurisdiction, trade-related goals and environmental concerns. With regard to these last, the following principles are of special interest: the principle that preventive action should be taken, the principle that environmental change as a priority should be rectified at source, the 'polluter pays' principle, the precautionary principle, the natural resource household principle, the substitution principle, the integrated pollution control principle and the principle of shared responsibility. As regards most of these principles, see Article 174, subsection 2, (ex Article 130r, subsection 2, Sands, 1995 (b)). The last principle does not indicate a departure from individual liabilities towards a notion of collective responsibility, merely participation by different interests, especially industry, consumers and policymakers. (Commission, 1997, p. 8)

The acceptance of the sustainability concept implied a serious challenge to the notions of Western law. The right of landowners to use natural resources must be regulated and supplemented by an obligation to the common good. Another challenge lies in the claim, typically stated in preambles on sustainability, that the range of factual considerations must be extended vis à vis the considerations which are traditionally found within a societal sector. This means that environmental principles and objectives may become integrated in sectors which were hitherto treated as autonomous. However, integration of environmental considerations has no radical transformation of the sectoral practice in mind, but is at best an external application of environmental principles, due to the fact that environmental issues themselves define a sectoral interest, giving them a competitive, rather than unifying or reconciling, function.

As far as the transport sector is concerned, the Treaty of Rome does not contain any explicit commitment to sustainable transport solutions. The focus in Articles 70-84, which are concerned with the transport sector, is on preventing discrimination and trade restrictions. However, on the action plan level, we find an elaboration of the possible meanings of sustainability in this area. In the light of European *environmental* policy, the main targets of sustainable mobility are: the securing of continuing access to natural resources, improving the quality of life, and the prevention of permanent environmental damage. To realise these goals, the planning of human mobility is of paramount importance. Mobility, interpreted in a purely quantitative, geometrical sense, supports and makes effective the establishment of an internal market, involving the free movement of labour, the expansion of the transportation network (motorways, high speed trains, etc.) and the liberalisation of freight transport.

The main dilemma in European transportation policies, and a major obstacle to sustainable solutions, is the dilemma between the internal market, which generates a considerable growth in transport, and consideration for environmental and social conditions, which demands a reduction in transport activities. This dilemma, however, is intended solved in two ways: *firstly*, by reducing ambitions, and at the same time setting out clear priorities, ensuring that the efficient actions undertaken are presumed to take place "under the best possible environmental and social conditions" (Commission, 1993, p.18). This indicates that the most important goal is to secure efficiency, and that the protection and management of the natural and social environment is expected to be secured under the conditions of an efficient functioning of the internal market. "The European Commission believes the best means of achieving [sustainable mobility] would be to create an integrated European network, based on the complementarity and inter-operability of modes of transport" (Bruyas, 1996, p. 2). *Secondly*, enhancement of the quality of life is looked upon as the free mobility of persons and goods, i.e. consumerism, but not as an optimal state of social life and natural environment. The basic policy is grounded in an economic rationality, stimulating economic growth, and at the same time demanding that the responsible polluters (actors) pay the costs of cleaning-up activities. Based on the "White Paper on Environmental Liability" (2000), the proposed EU-wide environmental liability regime should, however, cover both environmental and traditional damage to nature, especially to those natural resources that are important for the conservation of biological diversity in the Community.

Free mobility and liberalisation of trade have been decided on without ever raising the question of the subject and range of these transactions. Mobility for whom? Who and what is to be liberated and why? The lack of a qualitative discussion of human and non-human life conditions veils the strengths and weaknesses of the appeal for sustainability goals.

New Zealand's laissez-faire policy, and the attempt to integrate ecocentric values in environmental law

The New Zealand 'Resource Management Act' (RMA) from 1991 was one of the first national acts which aimed to secure sustainability in terms of "sustainable management of natural and physical resources" and which recognised the indigenous culture and the intrinsic value of natural phenomena. As such, it has been praised for its apparent ecocentric outlook. The credit for this evaluation must mainly be given to the attempt to integrate Maori values. All persons exercising functions and powers under the RMA are under an obligation, as a matter of national importance, "to recognise and provide" for the relationship of Maori to their ancestral lands, water, sites, waahi tapu (i.e. sacred places) and other taonga (i.e. treasures, including language and culture); to have "particular regard" to kaitiakitanga (i.e. guardianship of resources) and to "take into account" the principles of the Treaty of Waitangi. Furthermore, all local authorities, when preparing or changing plans or policies, are under an additional obligation to "consult" with tangata whenua (i.e. local natives or "people of the land") and to "have regard" to any relevant planning document of an iwi (i.e. tribe) authority (Boast, 1993, p. 248).

Yet, formally recognising these values – and with them a nearly eco-centric interpretation of sustainability – does not guarantee effectuation. While some legal disputes have been settled by redefining (and largely misinterpreting) Maori tenets, other have simply been excluded – as was the case with property rights. The main reason for avoiding the issue of property or land rights in the RMA was the intention of avoiding long-drawn-out conflicts between the historical interests of the "Pakeha" and the interests of the Maori. This was made possible by a recent change in emphasis from "land-use regulation" to the administration of "the adverse effects of activities on the environment". The question of moral and legal responsibility may thus be settled relative to the polluter and independent of the owner (although the latter will normally be an important actor).

As Maori values, even where formally recognised, are of minor importance, practically speaking, the consequences are that environmental considerations play a relatively inconspicuous role in the legal framework. The hierarchical structure of RMA in fact prescribes a subsumption of these considerations under the general socio-economic purposes as stated in Art. 5(2):

> "'sustainable management' means managing the use, development, and protection of natural and physical resources in a way, or at a rate, which enables people and communities to provide for their social, economic and cultural well being and for their health and safety while (a) sustaining the potential of natural and physical resources (excluding minerals) to meet the reasonably foreseeable needs of future generations; and (b) safeguarding the life-supporting capacity of air, water, soil, and ecosystems; and (c) avoiding, remedying, or mitigating any adverse effects of activities on the environment." (Art. 5(2))

This priority is made clear in "Transit New Zealand", the national transport policy plan, which is supposed to implement the sustainability claims of the RMA in the transport sector. Its purpose is

> "to promote policies and allocate resources to achieve a safe and efficient land transport system that maximises national economic and social benefits." (See Transit New Zealand, p. 6)

In order to secure a safe and efficient transport system, potential conflicts with Maori interests and environmental constraints must be taken into consideration. Ultimately, however, the goals are that the transport policy is to secure the development of the road network and the establishment of facilities to meet the growing need for efficient transportation and "a high level of service" in a safe and humane way. With the liberalisation of resource management ensured by the former zone-based planning, and the redundancy of the concept of incompatible land use, and with the subordination of Maori values under economic interests, the RMA ultimately provides "a laissez-faire environment in which people should be able to do whatever they want" (Oc, 1992, p. 331).

Japanese pragmatism, collective responsibility and sustainability

Before the enactment of the Basic Environmental Law (BEL) in 1993, Japanese environmental regulation was mainly determined by two general laws, the "Law on Environmental Pollution Control" or "Public Nuisance Countermeasures Basic Law" passed in 1967 to combat serious industrial pollution, and the Nature Conservation Law from 1972. The ambition was to stop the deterioration and destruction of outstanding features of the national environment. The overall purpose of these acts was to provide effective solutions to pressing problems and to identify responsibilities, which, with regard to industrial polluters, would result in the recognition of collective or co-operative responsibility. Entrepreneurs were responsible for co-operating with State and local authorities on matters of public nuisance, and residents had a duty to contribute to preventive actions. (EHS Law Bulletin, 1967, p. 2-3). Neither of these acts encompassed sustainability, but must be regarded as a pragmatic reaction to publicly perceived environmental deterioration.

The BEL is designed to work as a general interpretative framework for past and future environmental legislation, and contains explicit references to the idea of sustainability. Its main intentions have been summarised in three principles:

> "(a) The blessings of the environment should be enjoyed by the present generation and succeeded to the future generation. (b) A sustainable society should be created where environmental loads by human activities are minimised. (c) Japan should contribute actively to global environmental conservation through international co-operation." (Environmental Agency, 1995, p. 2)

Article 4 of the BEL specifies as a goal of the Act, that a:

> "sustainable development is ensured by fostering sound economic development with reduced environmental load, through practices on environmental conservation such as reducing as much as possible the environmental load generated by socio-economic and other activities, which are voluntarily and positively pursued by all the people sharing fair burden".

These ideas, which reflect the UNCED-report with respect to intra- and inter-generational equity, sound economic development, and globalisation, have been implemented far beyond conservation laws. The commitment to sustainability has resulted in a general action plan, "Basic Environmental Plan", enacted in December 1994. As an action-orientated interpretation of the BEL's insistence on sustainability, its verbal commitment to an intensification of "co-operation" and "fair burden sharing" and its emphasis on "sound material cycle" and a "harmonious coexistence between nature and human beings" (see Environmental Agency, 1995, p. 3), is very much in line with the international consensus on environmentally sustainable development.

However, there is apparently no trace of these intentions in specific sectoral policies, which is especially true for the transport sector. In Japan, the concept of "sustainable mobility", is scarcely used. Instead, the main emphasis is on a "comfortable society". In 1992 the long-term vision of road transport marks a shift from:

> "traditional vehicle-oriented road construction to 'human-orientated road construction'. It aims at the enrichment of living conditions, vitalisation of regional communities, and realisation of road environment which is considerate of both people and nature." (Japan Road Association, 1993, p. 18)

The phrase "considerate of both people and nature" is open to various interpretations. The strong emphasis on historical, cultural roots (see Japan Road Association, 1993) suggests two likely responses: *firstly*, the notion of "shared" (co-operative or collective) responsibility, as seen in connection with early environmental regulation, could also be seen as operating in transportation policies. Roads are looked on as public property, and there is a long legal tradition regarding the maintenance of roads as the responsibility of the local inhabitants (the people living "around" them). This fact still has an important impact on questions of legal responsibility and administrative regulation. Although the issue of inter-generational justice (equity) is hardly mentioned in official documents, the historically established practice of making "collectives" legally and morally responsible may nevertheless turn into a valuable condition for the implementation of sustainability demands for the sake of future generations. *Secondly*, the reluctance to deal with, or rather, lack of familiarity with, abstract concepts and principles invites a pragmatic, case-orientated handling of sustainability issues, rather than a theoretical clarification of premises and implications. In this respect, there is no general interpretation of phrases such as "considerate of people and nature".

ETHICAL RATIONALITIES USED BY LEGISLATORS IN THE CONCEPT OF SUSTAINABILITY

When varying historical and cultural conditions confront sustainability demands, serious obstacles to their unanimous global interpretation and implementation are revealed. Essentially, the divergent interpretations and the potential for conflicts lurking behind an apparent verbal agreement may be traced back to different ethical rationalities, legal traditions and socio-economic conditions (Barton, 1996, p. 119, Sarre, 1995, p. 125 & Thompson, 1991, pp. 243-244). In this article, our emphasis is on differences in conceptualised rationalities. The discussion will be structured with regard to the problem of *centrism*, and subsidiarily from the point of view of the significance of *theoretical* inquiries. As far as centrism is concerned, there is an important difference to be observed in both theory and practice between (a) an attitude which holds that man expresses and realises himself in nature, and (b) an attitude where nature is said to express itself in man (cf. Meyer-Abich, 1993). *The first*, being an attitude of command and suppression, control and regulation, is normally called an expression of anthropocentrism. *The second*, seeing man as being self-realized with nature, of which he is a part, can be labelled "ecological or ecocentric thinking".

Anthropocentrism's predicament

Ethical anthropocentrism is *prima facie* a reasonable position, if it is understood as the asymmetry of moral obligations. That is, animals and ecosystems may have rights (no necessary assumption), but only humans have obligations and responsibilities. In other words: human beings are the paradigmatic moral beings, both as 'moral agents' and 'moral discussants'. But when it comes to 'moral subjects', i.e. the question of who may be part of our moral concern, it is rather difficult to defend a human-centred standpoint.

As some critics of anthropocentrism have stated, we have no definitive reasons for positing an ethical line of demarcation between human beings and other animals, or even plants, insofar as there are goals which may be thwarted or opposed in a way which open-minded moral agents might find offensive (e.g. Taylor, 1986). In fact, from a philosophical point of view, any proposed criterion of demarcation, e.g. moral or intellectual capacity, biological complexity, suffering, or interests, is arbitrary, anthropomorphistic and in an important sense irrational (Zeitler, 1987, pp. 98-111).

So, in a non-trivial sense, in arguing in favour of the moral superiority of human beings as moral *subjects* on philosophical grounds (theoretical reflectivity), ethical anthropocentrism runs into serious problems. There might, however, be another rationality (bodily experience) in favour of human supremacy, based on socialised feelings and experiences supporting our perception of the task of human beings as agents with the moral right and physical power to subdue non-human nature. Yet, while this approach is a historical fact, its dependence on our anthropocentric civilisation is more an *explanation* of our anthropocentric thinking and feeling than a fact of (human) nature as such. In reality, a slightly different socialisation, strengthening our sensibility to "nature" and awareness of "environmental" problems, might lead to a radically different self-understanding within proper institutional frameworks. The more we rely on bodily (sensuous) perceptions and the less attention we pay to rational (logical) approaches, the more true this is, since those approaches are specific human 'instruments'.

Apart from these general problems, some have stressed that anthropocentrism is the foundation for our current environmental crisis, and therefore an impediment to any endeavour towards sustainability. Although different versions of anthropocentrism e.g., stewardship models, have lately been worked out in an effort to meet the present environmental challenge, it is difficult to see how they may work outside a Christian belief in Creation. More importantly, the concept of *moral autonomy* and its conceptual distinction from "natural" or empirical conditions, presupposed by any anthropocentric attitude, and central to modern European/North American culture, is hardly consistent with the fact of social and natural interdependence. At the very least, the dualism presupposed by anthropocentrists poses huge theoretical problems, which still lack an adequate conceptual solution. Certainly, our self-image as autonomous agents has had a significant negative impact on our actual unsustainable behaviour.

The interpretation of the objectives of the *European Community* is mainly based on an understanding of "nature" or the "environment" as *constraints for human development*. There is no suggestion at all that natural phenomena be given any moral status apart from their instrumental value to mankind. This attitude has its classic expression in the German philosopher Immanuel Kant's statement that the reason we should abstain from hitting our pets is because of the educational effect on human relations, not because of the animals' alleged moral status. Not only is a moral egalitarianism between different forms of life out of the question; there is no reason to select specific natural phenomena for moral recognition. Only aesthetical and instrumental arguments are to be used in dealing with the relationship with nature.

European Community environmental policy is caught in an anthropocentric institutional and mental framework. It ignores the systemic interdependence and potential intrinsic worthiness of all natural phenomena. Furthermore, its focus on the interests of the Member States amounts to ignoring the vast majority of the human population, thereby constituting an obstacle to the realisation of intra- and intergenerational justice.

What is lacking in Community legal documents, however, is partially expressed in a number of international statements to which the Member States have given their consent. For example, a draft of an "International Covenant on Environment and Development" expresses an ecocentric viewpoint by stating:

> "Nature as a whole warrants respect; every form of life is unique and is to be safeguarded independent of its value to humanity." (IUCN 1995, Art. 2, p. 31)

Other documents, recognising the intrinsic value of ecosystems and species, etc., include the treaties concerned with the Antarctic, the World Heritage Convention, the Berne Convention on the Conservation of European Wildlife and Natural Habitats, the CITES Convention, the World Charter on Nature, and several marine pollution agreements (Birnie & Boyle, 1994, pp. 193-194).

Intrinsic values and their appropriate place

Although anthropocentric elements are conspicuous in New Zealands' environmental and, especially, transport policy, different interpretations can also be put forward. According to Palmer, both anthropocentric and ecocentric interpretations are important, and must be balanced. The question is, however, whether such a balance is possible.

The interpretation of "sustainable development" chosen in the New Zealand RMA is heavily concerned with the social, economic, and cultural well being, and health and safety of human beings and communities. The connection of these socio-economic values to ecological values as expressed in Maori culture and the RMA's reference to the intrinsic values of ecosystems is dependent on the priorities built into the law. There are unequivocal elements favouring an anthropocentric or "minimalistic" interpretation in the preparatory work on the RMA.

This minimalistic interpretation speaks of the inclusion of "ecological limits" (Croning, 1989, p. 2, p. 4 and p. 6) and refrains as far as possible from normative discussions, which demand difficult and fragile social consensus. It can be argued that the legal framework for sustainable management is not concerned with ethically and ecologically right solutions, but with the avoidance of serious damage to man-nature relationships, which may be effectively prevented by a minimalistic law, respecting the historically attainable social consensus.

Whether a minimalistic, anthropocentric interpretation is compatible with ecocentric attitudes in a pluralistic sense depends on the exact definition of ecocentrism. If it is to be meaningful to use the word, clear, distinctive features must be identified. Ecocentrism must assume that humans cannot freely choose their natural status (autonomy, interdependence, etc.), nor can they freely choose their responsibility in relation to nature, or determine the intrinsic value of natural phenomena. In other words, *there is a paradigmatic difference* between an attitude conceiving of natural phenomena as resources to be consumed unless consumption causes problems (anthropocentrism), and an attitude which makes our consumption of natural phenomena dependent on good reasons based an epistemological and ethical humility toward their life conditions. Ecocentrism is normally obliged to intuitively or rationally administer a radical precautionary principle of this latter type, and the RMA is supposed to follow such an approach. Nevertheless, the precautionary principle is still not very visible in New Zealand law, and the principle has never been applied in its radical version.

In specific cases, however, ecocentric thinking and practice may very well have a chance of being implemented, as is the case with respect to fishing rights, where "Kawanatanga" (i.e. the Crown's authority or governorship) and "Rangatiratanga" (i.e., the Maori peoples' residual sovereignty) have been co-ordinated, allowing the Maori to manage their own resources. It has been argued that, "(i)n essence the Treaty requires that the relevant planning authority for Maori resources be Iwi [tribe members/authorities] themselves" (Barns, 1988, sect. 2400). But, if Maori self-management threatens the overriding objective of nature conservation and protection, the Crown still has the power to intervene.

To the extent to which Maori thinking can be assumed to be effective, the focus on intergenerational justice can be given remarkably clear expression. (Bosselmann, 1995, p. 130). While Western academics still search for a consistent, clear theory and understanding of the obligation to tend to sustainability with a view to protecting the interests of future generations, the Maori have a very long, strong tradition for such concern, based on the concept of collective responsibility. This concept implies that a collective persists over time, and is not dependent on the existence of particular individuals. As a consequence, collective responsibility involves ancestors and future generations as much as those presently alive. While accountability and duty transcend the living, the implementation of this duty and responsibility will always rest with the living members (Perrett, 1992, p. 30).

Now, this conceptualisation may be very valuable, but it is essentially dependent on the idea of collective responsibility, which conflicts with the European concept of individual responsibility. However, the challenges of forging international environmental law have now smoothed the way towards a concept of responsibility which transcends the individual actor. As Hohmann has observed:

> "The recognition that all – even distant – states can simultaneously be violaters and victims with respect to accumulated pollution paved the

> way (together with the development of new prevention and monitoring technologies) for the gradual relinquishment of full proofs of causality (concerning abstract scientific causality and individual cause-effect relations)." (Hohmann, 1994, p. 187)

But with these qualifications, the current legal system in the West is hardly fit for genuine Maori concepts. There is no doubt that the verbal integration of Maori concepts in the RMA (and other legal acts) has had an ideological function, designed to pacify Maori claims and avoid social unrest. Moana Jackson from the Maori Legal Service in Wellington puts it clearly and sharply: the Act

> "captures, redefines, and uses Maori concepts to freeze Maori cultural and political expressions within parameters acceptable to the state."

In other words,

> "while Pakeha politicians no longer reject a notion of Maori rights, they see it sourced in their authority through biculturalism, rather than in Maori authority through rangatiratanga." (Jackson, 1992, p. 8-9)

Beyond anthropocentrism and ecocentrism

While the case of New Zealand illustrates the possible conflicts between two paradigmatic rationalities – anthropocentrism and ecocentrism – there are cases where this distinction seems to give no meaning at all. Japan may thus be interpreted as being *beyond* anthropocentrism and ecocentrism. Remaining for a moment in the language of "centrisms", the BEL is clearly anthropocentrically oriented, its general purpose (Article 1) being the promotion of:

> "policies for environmental conservation to ensure healthy and cultured living for both the present and future generations of the nation as well as to contribute to the welfare of mankind, through articulating the basic principles, clarifying the responsibilities of the State, local governments, corporations and citizens, and prescribing the basic policy considerations for environmental conservation."

Nevertheless, given this human-centred outlook, commentators have repeatedly emphasised that the Act builds on a respect for the intrinsic qualities of nature by using expressions such as "enjoy the *blessings* of a healthy and productive environment" (Article 3), which gives natural phenomena the status – not just of exploitable resources – but of a "gift" and largely independent source of value. Article 14 of the BEL stipulates that environmental policy should be concerned with a triple purpose:

> "to protect human health, to conserve the living environment and to properly preserve the natural environment".

The protection of biodiversity and the conservation of "various features of natural environment" must proceed "in accordance with the natural and social conditions of the area". Furthermore, it is an important provision "to maintain rich and harmonious contacts between people and nature" (Imura, 1995, p. 9). These specifications clearly indicate, while recognising the central importance of human agents, for whom this Act has been established, that both the living and non-living environment have a legal status making them entitled to human protection and preservation.

With the introduction of "Environmental Quality Standards" (EQS) – well known from the American NEPA – a precautionary principle has also been introduced, an anthropocentric interpretation of which seems unavoidable.

> "The EQS of Japan is prescribed in view of a sufficient safety margin so that there would be no adverse effects on the health of the general public from a medical standpoint, and although pollution may in certain cases slightly exceed the value of the standard, it will not immediately threaten the health of people." (Environment Agency: Air Pollution. Tokyo 1994, p. 3)

However, it might be problematic to propose an interpretation in an anthropocentrism-ecocentrism-framework. At the level of theoretical reflectivity, the applying of a Western framework of rationality to pragmatic attitudes implicit in Japanese environmental politics and practice might even be described as inconsistent or paradoxical. This can be clearly observed in daily behaviour. The zealous, careful concern for a small potted plant squeezed into the wall, or the attentive study of a single "sakura" (cherry blossom) coexist naturally with the disregard for unrelated community yards, the neglect of river conditions or of the global environment. Heedless waste-disposals and insensitive development projects are likely to overshadow the deep concern for specific natural beings in a particular life context. While specific natural phenomena, related in some way to daily activities, may be given close, very intensive attention; nature conservation issues, animal rights movements and philosophies of nature remain abstract, external, irrelevant issues.

Correspondingly, an analysis of the Japanese language discloses two important things: *firstly*, the dependence of the choice of appropriate terms and concepts on the character of the specific situation in a way which presupposes an intuitive understanding of, and familiarity with, its circumstances. *Secondly*, and closely connected, there is the basically non-dualistic attitude imbuing linguistic understanding. Seeing man, not as an autonomous individual, but as living out of the "between man and man" is an example of a language structure which moves beyond the subject-object dichotomy. The subject-free nature of Japanese language is reflected in a corresponding non-individualistic morality (Kimura, 1995, p. 14) and in the attitude towards technology.

Technology is for many Japanese a natural, instrumental part of their lives. Life is natural in such a way that they would not acknowledge that "nature" might be unable to deal with waste and pollution. Anything fits into the pragmatic spirit of society: technology, different religions, different life-styles, as long as it enriches their lives.

The attitude to and solution of environmental issues is not dependent on the development of theoretical models, but on the specific circumstances of action (cf. Döbert, 1997, pp. 99-101). The sustainability of these attitudes and actions is, however, conditional on a non-dualistic practice confident of basic familiarity with the specific qualifications for symbiotic behaviour. But the stability of these conditions is under severe attack in modern Japanese society. Consequently, Western modelling of environmental research, including environmental ethics, is gaining a foothold. To a degree, at least, the last hundred years of westernisation has caused an imbalance in the expression of Japanese society. So, perhaps Philip Sarre is right when he says that Eastern values and attitudes – although consistent with environmentalism – "are strongest in stable local communities" (Sarre, 1995, p. 122).

CONCLUSIONS

The prevalent opinion is that sustainability only makes sense on a global scale. This does not imply that action towards sustainability has to take the form of joint global action on the basis of a shared global understanding. On the contrary, as the foregoing comparative analysis has illustrated, there are paradigmatic differences between cultures and traditions as to the perception of "nature", "environmental problems" and "quality of life". To try to enforce a global consensus on these issues is not only morally reprehensible, but also unrealistic and inefficient. Thus, a more appropriate way to deal with environmental, social and economic problems would be to base global policy on respect for local traditions and culturally determined rationalities and perceptions. Correspondingly, reflecting the variety of legal systems, incorporation of the concept of sustainability will be carried out in accordance with the legal tradition of each state (Basse 1997, p. 21). They may all contribute to more sustainable solutions, although in different ways.

Ecological movements have propagated the idea that we should "think globally and act locally". The analysis showed that this formula might be in need of revision. There are certain limits to the potentiality of *thinking globally,* although continued westernisation, the internalisation of the market economy, and the formation of global organisations might eventually lead us to this condition as an empirical fact. However, under present conditions, the new ecological formula must treat the local and global consequences of thinking and acting locally, presupposing that only on a local level might we have a chance of truly being integrated with nature as the precondition of sustainable development, and that only in local interpretation does it make sense to relate to global conditions.

We need no universal morality, states Klaus Michael Meyer-Abich – it would suffice that we not violate our own morality permanently and systematically by living at the expense of others – the Third World, posterity and the connatural world. (Meyer-Abich, 1997, p.210, cf. ibid., pp. 211-215, Weber-Schäfer, 1997 & Fuchs 1997).

Yet, one might reply, it might be said that present environmental conditions are so serious, globally and locally, that this decentralised approach would be quite dangerous. Aggregate impacts are difficult to detect. Environmental problems are trans-boundary problems, and one's responsibility may not be mainly an individual responsibility, but collective in nature, (cf. the "Principle of Non-Discrimination",

Hohmann, 1994, p. 312 and the concept of "Responsibility Without Fault", Yamagami & Suketake 1994, pp. 40-41). The identification with one's "sphere of responsibility", although still locally delimited, must not exclude openness to what is going on beyond the place or sphere within which one is supposed to act. To some extent, and especially in critical situations, the need for instant action on a global basis demands joint agreements and joint actions across national and cultural borders. It is thus an important feature of international environmental law to think "in terms of joint responsibility, regional co-operation or global common concern criteria" (Hohmann, 1994, p. 300, cf. p. 311).

Still, although joint global action could be necessary, responsibilities are highly differentiated. In particular, it has often been emphasised that the North is required to take far stricter measures than the South, and that it should provide special assistance due to its excessive level of consumption (see e.g. Ebbeson, 1995, p. 223).

The emphasis on local thinking and action is not only conditioned by our ability to be familiar with the substance and effects of our particular actions and policies. It also directs our focus away from goal-oriented, measure-based strategies to a *procedural* concept of sustainability (Birnie & Boyle, 1994, pp. 194-195). What is important in order to find a reasonable balance between human and non-human ("environmental") concerns is not the attainment of some agreed-on ideal "state of nature", but the way one *responsibly* deals with daily challenges (Zeitler, 1999, p. 11). Correspondingly, an emphasis on the legal and political processes and instrumentality issues is required.

REFERENCES

Anker, Helle T. & Ellen Margrethe Basse (ed.) (2000). *Land Use and Nature Protection. Emerging Legal Aspects.* Copenhagen: Djøf Publishing.

Barns, Mike (1988). *Resource Management Law Reform. A Treaty Based Model – The Principle of Active Protection.* Working Paper no.27. Wellington: Ministry of the Environment.

Barton, Harry (1996). The Isle of Harris Superquarry: Concepts of the Environment and Sustainability. *Environmental Values,* Vol. 5, pp. 97-122.

Basse, Ellen Margrethe (1997). Regulatory Chain – Results of an International Development. In: Ellen Margrethe Basse (ed.), *Environment-*

al Law. From International to National Law, pp. 9-52. Copenhagen: GadJura.

Bengoetxea, Joxerramon (1993). *The Legal Reasoning of the European Court of Justice*. Oxford: Clarendon Press.

Birnie, Patricia & Boyle, Alan (1994). *International Law and the Environment*. Oxford: Clarendon Press.

Boast, Richard (1993). Treaty of Waitangi and Environmental Law. In: Christopher Milne (ed.), *Handbook of Environmental Law. Royal Forest and Bird Protection Society of New Zealand*, pp. 246-253. Wellington.

Bosselmann, Klaus & Taylor, Prue, (1995). The New Zealand law and conservation. *Pacific Conservation Biology*, Vol. 2.

Bruyas, Olivia (1996). *Transport and the Environment. European Provisions. Under the standard of sustainable mobility*. EIS.

Calliess, Christian (1997). Towards a European Constitutional Law. *European Environmental Law Review*, April, pp. 113-119.

Chiba, Masaji (1995). Legal Pluralism in Mind – A Non-Western View. In: Hanne Petersen & Henrik Zahle (eds.), *Legal Polycentricity: Consequences of Pluralism in Law*, pp. 71-84. Dartmouth: Aldershot.

Commission of the European Communities (1993). The future development of the common transport policy. *Bulletin of the European Communities*, Supplement 3/93.

Commission of the European Communities (1997). *Statements on Sustainable Development*. Luxemburg.

Commission of the European Community (2000). White Paper on Environmental Liability, *COM(2000)66 final.*

Croning, Karen (1989). The Intrinsic Value of Ecosystems. *Resource Management Law Reform*, Sustainability, Intrinsic Values and the Needs for Future Generations. Wellington: Ministry for the Environment, (Working Paper No. 24).

Döbert, Rainer (1997). Welche Wertsysteme/Weltbilder überleben den diskursiven Test? In: W.Lütterfelds, Th.Mohrs (eds.), *Eine Welt – eine Moral*, pp.77-103. Darmstadt: Wissenschaftliche Buchgesellschaft.

Ebbeson, Jonas (1995). *Compatibility of International and National Environmental Law*. Uppala: Uppsala University.

EHS Bulletin Series (1967). Public Nuisance Countermeasures Basic Law. Law No.132, Aug.3. *EHS Law Bulletin Series*, Vol. VII, YA.

Environmental Agency (1995). *Environmental Protection Policy in Japan*. Tokyo.

Fuchs, Martin (1997). Universalität der Kultur. In: Manfred Brocker &

Heino H. Nau (eds.), *Ethnozentrismus*, p. 141-152. Darmstadt: Wissenschaftliche Buchgesellschaft.

Government of Japan (1993). *The Basic Environmental Law* (BEL). Law No.91 of 1993. Tokyo.

Harris, B.V. (1993). Sustainable management as an express purpose of environmental legislation: The New Zealand Attempt. *Otago Law Review 541*.

Hohmann, Harald (1994). *Precautionary Legal Duties and Principles of Modern International Environmental Law*. London: Kluwer.

Imura, Hidefumi (1995). *National Environmental Policies. A Comparative Study of Capacity Building: The Case of Japan*. Nov. 7.

IUCN: Draft (1995). *International Convention on Environment and Development*. Gland.

Jackson, Moana (1992). The Colonization of Maori Philosophy". In: Graham Oddie and Roy Perrett (eds.), *Justice, Ethics, and New Zealand Society*, pp. 1-10. Auckland: Oxford University Press.

Japan Road Association, MICHI (1993). *Roads in Japan 1993*. Tokyo.

Krämer, Ludwig (2000). *EC Environmental Law*. Fourth Edition. London: Sweet & Maxweel.

Kimura, Bin (1995). *Zwischen Mensch und Mensch. Strukturen japanischer Subjektivität*. Darmstadt: Wissenschaftliche Buchgesellschaft.

Merryman, John et al. (1994). *The Civil Law Tradition – Europe, Latin America, and East Asia*. The Michie Company.

Meyer-Abich, K.M (1993). *Revolution for Nature*. Cambridge/Denton: The White Horse Press & University of North Texas Press.

Meyer-Abich, K.M (1997). Ganzheit der Welt ist besser als Einheit – Wider den Universalismus. In: Wilhelm Lütterfelds & Thomas Mohrs (eds.), *Eine Welt – eine Moral?*, pp. 203-216. Darmstadt: Wissenschaftliche Buchgesellschaft.

New Zealand Government (1994). Resource Management Act 1991. Wellington

Oc, Taner (1992). Performance-guided Planning – New Zealand's Resource Management Act. In: *Town & Country Planning*, Nov.-Dec. 1992, pp. 328-331.

Ono, Susumo (1996). *Nihongo no nenrin* [The annular ring of Japan], Tokyo.

Perrett, Roy (1992). Individualism, Justice and the Maori View of the Self. In: Graham Oddie & Roy Perrett (eds.), Justice, Ethics and New Zealand Society, pp. 27-40. Auckland: Oxford University Press.

Rehbinder, Eckhard and Richard Stewart (1988). *Environmental Protection Policy. Legal Integration in the United States and the European Community*. De Gruyter.
Sands, Philippe (1995). *Principles of International Environmental Law*, Vol. 1. Manchester: Manchester University Press.
Sands, Philippe (1995). International Law in the Field of Sustainable Development: Emerging Legal Principles. In: Winfried Lang (ed.), *Sustainable Development and International Law*, pp. 53-66. London: Kluwer.
Sarre, Philip (1995). Towards Global Environmental Values. Lessons from Western and Eastern Experience. *Environmental Values 4*, pp. 115-127.
Sharp, Andrew (1990). *Justice and the Maori. Auckland.* Oxford University Press.
Taylor, Paul W. (1986). *Respect for Nature*. Princeton: Princeton University Press.
Thompson, Michael (1991). Plural Rationalities: The Rudiments of a Practical Science of the Inchoate. In: Hansen, Jens Aage (ed.), *Environmental Concerns*, pp. 243-256. London: Elsevier.
Transit New Zealand (1994). *Transit New Zealand and the Environment.* Wellington.
Weber-Schäfer, Peter (1997). Eurozentrismus contral Universalismus. In: Manfred Brocker & Heino H. Nau (eds.), *Ethnozentrismus*, pp. 241-155. Darmstadt: Wissenschaftliche Buchgesellschaft.
Yamagami, Kenichi & Suketake, Eijun (1994). Japan's Constitution and Civil Law. Tokyo: Foreign Press Center.
Zeitler, Ulli (1987). Über die Grundlagen einer naturalistischen Ethik als Umweltethik. *Danish Yearbook of Philosophy*, Vol. 24, pp. 97-121.
Zeitler, Ulli (1999). *Grundlagen der Verkehrsethik*. Berlin: Logos-Verlag.

CHAPTER 5

Rationality Deficits in Behavioural Intervention Strategies

Erik Kloppenborg Madsen and Folke Ölander

INTRODUCTION

THIS PAPER AIMS to initiate a critical discussion of some core assumptions behind intervention strategies aimed at getting households to show more environment-friendly and, in the long run, more sustainable behaviour as buyers and users of goods and services. In order to do so, one must scrutinize basic assumptions concerning human motivation and decision-making, since different intervention strategies are explicitly or implicitly based on different assumptions concerning reasoning and motivation in human action.

By the expression "rationality deficits in behavioural intervention" we primarily refer to the lack of legitimacy and wider action orientation inherent in much regulation based on the instrumental rationality model. Deficits which may cause less commitment, less involvement, less responsibility and less self-confidence when it comes to people's perceptions of their own abilities. By the phrase we also intend to pinpoint change strategies which fail to allow for initiative and self-transformation, and which fail to promote and strengthen civic society. In our view, rather than designing strategies for the manipulation of individual behaviour which do not contribute to the development of competent coping abilities, one should search for strategies which aim to develop and strengthen individual as well as collective abilities and initiatives. That is, strategies which focus on the development of competent, responsible and involved citizens (rather than clients).

It is obvious that both the choice of a generic intervention strategy and the selection of a more specific method to be used in a particular context are determined, at least in part, by a preconception of how people think and act when confronted with complex environmental problems.

Three generic intervention strategies are often distinguished: legal regulation, the use of economic incentives and disincentives, and informational measures (often including persuasive and/or motivational elements). In relation to that classification, our focus will be placed on discussing the rationale behind, and the problems with, campaigns with an informational and persuasive aim. However, for our purpose, the most useful distinction is one that centres on whether a strategy focuses (a) narrowly on affecting overt behaviour and improving its effectiveness or (b) on equipping the individual citizen with wider coping abilities, while at the same time enhancing opportunities of collective problem-solving.

The paper attempts to point to the problems that will arise if, when selecting and designing intervention strategies, one assumes that individuals (here consumers or citizens) function according to the utilitarian means-end scheme in cases where their behaviour is better described by a norm or a rule-following scheme.

We see two major problems with intervention strategies based on the economic or utilitarian model.

1. An intervention may simply be ineffective if based on the wrong assumptions as to the individual's motivational structure. In the worst case, if a strategy is based on the wrong assumptions about people's self-perceived identity, there can be a counter-reaction, so that the effects turn out to be the opposite from what was intended (reactance, Brehm, 1966, or other forms for resistance, Eagly & Chaiken, 1993, Ch. 12).

2. An intervention based on the utilitarian rationality paradigm may be effective in the sense that it produces the behavioural changes that were desired. However, the means used can have the unintended side effect that they diminish the individual's personal responsibility, assuming – as we do – that reasoning and arguing as a way of life constitute a fundamental aspect of that responsibility. In the long run, there can be serious repercussions for a civic society that wants to build on reasoning, interpersonal communication, and other democratic virtues.

In this paper, we limit ourselves to discussing the second – and, one could argue, the most fundamental – of the two problems.

THE CONCEPTION OF HUMAN BEHAVIOUR IN PERSUASION STRATEGIES BASED ON ATTITUDE THEORY

When economic incentives or disincentives are chosen, the underlying assumption is that of individual instrumental rationality. Furthermore, one presumes that pecuniary motives play an important role for action (or lack of action) in this behavioural sphere.

The use of persuasion strategies usually builds on a conception of human behaviour similar to the economic utilitarian approach. Hence, these strategies more or less replicate the concept of rationality found in economics, and build on premises which echoing the economic view of the human character as defined by self-interest, and behaviour as utility maximizing. Typically, the persuasive approach is based on a general model of behaviour, in which attitudes – based on the actor's evaluative assessment of possible outcomes of a given act – are seen as a major determinant of the choice of act. The most well-known, and often cited, model of this kind has – in several consecutive versions – been advanced by Fishbein and Ajzen (Fishbein & Ajzen, 1975; Ajzen & Fishbein, 1980; Ajzen, 1985, 1988, 1991).

The theory of reasoned action suggested by Fishbein and Ajzen (1975) is based on the idea that it is possible to give an account of human behaviour merely by referring to a few concepts defined in a theoretical frame of reference.

The core assumption is that people act rationally in an instrumental sense and that they apply accessible information in a systematic way. Unconscious motives and other overpowering determinants are not supposed to play a role. The intention to act is based on deliberation, and action is determined by the behavioural intention. The congruence of action-intention and action may be interpreted as a hardcore model assumption (using Lakatos' terminology, 1970) in the Fishbein/Ajzen model, meaning that if there is an observed difference, this is necessarily due to the influence of other factors, such as physical barriers or unforeseen events. Such factors do not belong to the model, however, and later versions introducing additional explanatory variables, such as "perceived control" (Ajzen, 1985, 1988, 1991), still build on the hardcore assumption of congruence between intention and behaviour, albeit a person low in perceived control will hold a weaker behavioural intention.

The theory of reasoned action is thus based on a model which assumes that people follow an instrumental or – as phrased by March & Olsen (1989, 1995) – a consequential logic. It is also a model which, using the notion of society as constituted of system and life world (Habermas, 1981), does not take into account the "horizon of meaning" which is built into action orientations in a life-worldly perspective, as opposed to a systemic perspective. And it is therefore a model which proceeds from a conception of individuals as intelligent bundles of needs and wants, rather than as culturally moulded identities.

Usually, consequences other than pecuniary ones are assumed to be taken into account, and there are cognitive limits to what the individual is able to include in his calculations (in terms of available alternative actions and their perceived consequences); hence the theory of reasoned action can be regarded as a version of the subjective expected-utility model often put forward by students of decision-making. It can be described as an example of a theory of "bounded rationality" (Simon, 1956, 1972). But since it is based on the idea that people attempt to maximize their personal utility, it is basically an individualistic model. Social norms enter into the model in some of its versions, but then only as one of the dimensions taken into account by the individual in his "calculating" behaviour.

Later expansions of this model by other researchers (Aarts, Verplanken, & van Knippenberg, 1998; Grunert & Juhl, 1995; Pieters, 1991; Thøgersen, 1994; Triandis, 1977; for a review of many of them, see Conner & Armitage, 1998) have added various factors: habits, knowledge, situational factors and values (which can influence the weight given to various outcome dimensions). These expansions do not, however, alter the basic structure of the model or the rationality assumptions which are an inherent part thereof. One could perhaps say that the added variables provide a protective belt of auxiliary hypotheses around the hardcore assumptions. Generally, these models all have in common that the individual considers the consequences of her or his potential actions, basing the choice of an action on a weighing of the advantages and disadvantages to her- or himself linked to these consequences.

The strategic implications of the theory of reasoned action are quite straightforward. Taking for granted that the major determinant of people's behaviour is their behavioural intention, the important question for the regulator or persuader concerns the determinants of a person's behavioural intention. And according to this theory, the intention to behave in a given manner is a function of two factors: the individual's attitude toward the behaviour in question, and the individual's subjective norm (what the individual thinks others want her/him to do). A change in a person's intention to behave in a given manner can therefore be brought about if either the attitudinal component or the normative component, or the relative weighting of the two components, is changed (O'Keefe, 1990).

THE CONCEPT OF RATIONALITY AND THE AIMS OF REGULATION

Social regulation, defined as governmental programmes initiated in order to protect the quality of the environment, may be seen as endeavours of societal self-improvement, reflecting more or less of a consensus concerning the goal or object of regulation (Sagoff, 1988). Few would question the legitimacy of the aim of improving the quality of the environment. Much dissent can be expected, however, as to the means that should be used in order to reach this goal. There will also be considerable disagreement about more specific definitions of environmental problems.

From a more comprehensive viewpoint, the extent to which the aims of social regulation can be reached may be seen as basically depending on (a) the possibility of reconciling collective and individual values, beliefs, and interests, and (b) the possibility of successful co-ordination of human action. As a minimum, some reconciliation and some co-ordination of action are imperative if modern societies are to meet the environmental challenge facing us all.

This is a challenge that places a burden on accounts of reason and rationality as individualistic and self-regarding. One can even argue that modern instrumental rationality is the origin of much environmental degradation (see Beck, 1994): directly, as the dominating mode of reasoning, and indirectly, as defining norms, roles, identities, ways of life, civic responsibilities and commitments, etc. The realization that instrumental rationality itself plays a crucial role in the creation and maintenance of environmental problems does not, however, justify the all-encompassing kind of rationality criticism put forward by some students of the role of reason in modern society (see the discussion in Szerszynski, 1996). The statement that "practices do not exist without a certain regime of rationality" (Foucault, 1987, p.107), is meant to express the idea that specific forms of rationality are inscribed or institutionalised in *every single* practice. Even when one acknowledges the merits of research which focuses on rationality as historically situated, it is difficult to see how modern societies should be able to meet the environmental challenge without some kinds of reasonable strategies. Rather, environmental issues challenge human rationality in two ways: by questioning instrumental rationality as the overarching norm, and by asking for another form of rationality without its flaws

When discussing the notion of rationality and its use in social regulation, one must therefore, it might be argued with Putnam (1987), lose sight neither of the immanence of reason, nor of its transcendence. That is, never see rationality decontextualized, but always in relation to cultural context and societal institutions. Moreover, rationality is coupled not only to established societal institutions, such as markets or administrative functions; rationality notions can also be traced in those cultural dimensions of society related to roles, identities, and perceived human responsibilities. At the same time, however, one has to recognize the need for a normative concept, a concept that expresses a universal regulative principle.

Whereas there is thus an inherent tension between rationality as an institutionalised mode of conduct, and rationality as a regulative idea, it is clear that in most cases of social intervention, the latter view is the prevailing one, and that the dominating concept of rationality is that of instrumental rationality. Instrumental intervention is believed to be rational in two respects. Firstly, it is rational in the sense that certain instruments of behavioural change are believed to be the most effective or efficient in reaching certain pre-established goals. Secondly, rationality is believed to characterize a certain cognitive and motivational structure laid down in the target population. This means that rationality is more or less thought of as identical with efficiency, and that the actors are more or less believed to be self-serving utility maximizers.

TWO PERSPECTIVES ON SOCIAL REGULATION

Bellah, Madsen, Sullivan, Swidler, and Tipton (1985) took up an important theme by pointing to the tension between citizenship and "professional rationality," taking their point of departure in Tocqueville's idea that "administrative despotism" would bring about in modern society a "form of government that erects over its citizens an immense protective power which is alone responsible for securing their enjoyment and watching over their fate." Such governance, Tocqueville argued, "does not break men's will, but softens, bends, and guides it; it seldom enjoins, but often inhibits action; it does not destroy anything, but prevents much from being born; it is not at all tyrannical but it hinders, restrains, enervates, stifles, and stultifies". "I do not expect their leaders to be tyrants, but rather schoolmasters." "Under this system the citizens quit their state of dependence just long enough to choose their masters and then fall back into it." (Tocqueville, 1835/1978, as quoted by Bellah et al., 1985, p. 209).

According to Bellah et al., professional managers and experts "may become the benevolent schoolmasters of Tocqueville's administrative despotism." The problem is that the professional preoccupation with effectiveness and efficiency of regulation does not nurture individual responsibility and citizenship. The professional point of view "tends to assume the validity of a trade-off between utilitarian efficiency in work and individual expressive freedom within private lifestyle enclaves" (Bellah et al., 1985, p. 210).

In order to be able to challenge the utilitarian assumptions about people's motivational make-up, one obviously needs another version of rationality and motivation, a version which may also lead to other views on the regulation of behaviour.

The "professional rationality" concept may be related to the Hobbesian view of society (1651/1968) whereas the argument emphasizing individual responsibility and citizenship – which we shall adopt – is more in line with the tradition emanating from Rousseau (1762/1987). The core differences concern (a) the assumptions that are made about human reason and motivation, (b) the relationship between individuals and collectivities, and finally (c) the question of whether motivational assumptions and beliefs may not themselves be reinforced by means of social intervention (i.e., a certain framing effect produced by the method of intervention itself.)

Re (a). Compared to Hobbes, Rousseau represents the more sophisticated version of human nature. Whereas Hobbes saw human nature as rooted in a pre-social setting, resulting in an unchangeable motivational structure, Rousseau believed that the transition from nature to society also implies a transition from instinctual action to moral action. Hobbes sees only "the individual in society" whereas Rousseau sees in addition "society in the individual."

It is this difference in outlook which shows itself in the ongoing argument between the models of economic man and social man, respectively. Or between self-serving utility maximizers and institutionalised role players. Or between *homo oeconomicus* and *homo sociologicus*. Or between action based on consequentialism and action based on appropriateness.

Hobbes and Rousseau's conceptions of human nature also differ in their implications when it comes to such issues as social regulation. Regulation by economic incentives more strongly echoes the Hobbesian point of view, whereas regulation by enlightenment strategies (information, education, learning) basically echoes the views of Rousseau.

Re (b). As to the relation between the individual and the collectivity, the two competing notions differ in the way joint goals are formulated. According to the Hobbesian view, individual desires are given and unalterable, and the function of the state or regulative authority is to aggregate individual needs and wants. According to Rousseau's view there is an important difference between "the general will" and "the will of all." It is the general will, and not the aggregated "will of all," which plays the core role in Rousseau's conception of society.

Re (c). From the Hobbesian point of view, intervention does not qualitatively change the individual or the social order itself. Intervention changes only behaviour, not motivational structure or identities, because human nature itself is unchangeable, and the social order is only altered quantitatively, in the sense that the state may be more or less firm in its control of citizen behaviour.

Rousseau's view differs, because intervention creates institutions, and institutions create roles, norms, and identities. According to this view, intervention by economic means not only builds on the notion of the *homo oeconomicus* but also reinforces *homo oeconomicus* as role and identity. Intervention by information and communication, on the other hand, builds on and reinforces participation and the identity and role of *homo politicus* (or at least has the potential to do so).

Some of these problems have been apparent for at least a century in the tension between sociologically and economically founded guidelines for conduct. Do people base their actions on norms (values or substantial rationality) or do they base them on calculations of the expected utility of actions (instrumental or strategical rationality)? The Parsonian criticism of the utilitarian (and positivistic) theory of rational action (the intrinsic means-end scheme) contains at least one important observation (Parsons, 1968). The intrinsic conception of rationality not only involves the elements of ends, means, and situational conditions as prerequisites for action, but in addition also the norm of the intrinsic means-end relationship itself (Parsons, 1968, pp. 698-699). In fact, the action scheme has as its premise that ends are not chosen according to the logic of the action scheme itself. They are given, and therefore not discernable from the element called "conditions of the situation." As stated by Østerberg (1980, p. 65), the ends become part of the conditions of action. Thus, the utilitarian scheme of action presupposes that there is at least one other norm than that of instrumentality. The final transgression of the instrumental theory of action lies in seeing it as one normative orientation among several others (Østerberg, 1980, p. 66).

FROM NEEDS AND PREFERENCES TO IDENTITIES AND MEANING

Based on the works of Habermas (1981, 1992) and March and Olsen (1989, 1995), as well as the perspectives on social regulation suggested by Sagoff (1988), one could adopt the view that behaviour is very often contingent on norms and roles, and the person's conception of "appropriate" behaviour, and assume that individual action depends on the answers to questions such as these (March & Olsen, 1995, p. 7): "What kind of person am I? What kind of situation is this? What does a person such as I do in a situation such as this?" As stated by March and Olsen, this is a view of human action as impelled less by anticipation and evaluation of its uncertain consequences than by a logic of appropriateness, reflected in a structure of rules and conceptions of identities. This model does not deny that self-interested calculation takes place, but such calculation is seen here as simply one of many rules that may be socially legitimised under certain circumstances. Furthermore, definitions of alternatives, consequences, and evaluations are strongly affected by the institutional context in which actors find themselves.

In this view, human action is the expression of "what is appropriate, exemplary, natural, or acceptable behaviour according to the (internalized) purposes, codes of rights and duties, practices, methods, and techniques of a constituent group and of a self." People act from understandings of what is essential, from self-conceptions and conceptions of society, and from images of proper behaviour (March & Olsen, 1995, pp. 31-32).

When individual households and consumers are confronted with a choice between various actions in which some lead to environmentally more beneficial consequences than others, it is often the case that from a purely egotistical perspective, the environmentally less beneficial consequence is clearly to be preferred (less expense, fewer personal disadvantages). Hence, most observed choices of environmentally more beneficial actions, and most attempts to bring about the choice of such alternatives by a "third party" such as a government, will – in the "consequential logic model" – have to assume, or to be able to induce, a preference "for the common good."

In the "institutional model," on the other hand, such action can be explained by reference to "rule-following," that is, a matching by the individual of an identity to a situation. In such a case, a choice of the environmentally more beneficial alternative will be based on the fact that it is "appropriate" to behave in that particular way in that particular situation. It is not then so much a question of the individual choosing between private and public gains, but between two "logics": a preference-based consequential logic and an identity-based logic of appropriateness (March & Olsen, 1995, p. 39). This said, one must realize, of course, that it by no means follows that the individual will always regard the environmentally beneficial action as the most appropriate!

We would like to argue, in line with the above-mentioned writers, for a different view on social intervention or regulation. This involves a transformation of the notion of rationality as instrumental and exclusively linked to biological and individual needs and wants, into a notion stressing cultural context, roles, and identities. It also stresses the need for the development of both individual action competence and collective civic competence. That is, the primary concern is not short-run efficiency but rather the capacity for responsible long run problem-solving by households, citizens, and local communities alike. Note that the notion of action as based on a "logic of appropriateness" associated with a certain practice does not, in our view, exclude such cognitive phenomena as reasoning and deliberation. Rather than seeing practice as constituted by rules of thumb, the matching of behaviours and situations may include a broad range of rules – from routines to reflective and deliberate problem-solving.

To support our position, we will need to briefly discuss how the task of social regulation can be based on assumptions about the individual and the society differing from the self-serving maximizer suggested by rational choice theory.

DEMOCRATIC GOVERNANCE IN AN INSTITUTIONAL PERSPECTIVE

The view of social intervention presented by March and Olsen is based on the distinction between two perspectives on governance (1995, p. 7). These are the exchange perspective, which assumes that the individual acts as a *homo oeconomicus*, i.e., as a calculating maximizer, and the institutional perspective, which assumes that, in acting, a person expresses identities and fills out roles in the endeavour to match situations.

The commitment to rationality inherent in the exchange perspective, understood as the belief that justification of action must be based on an evaluation of its consequences for the actor, is a widespread tenet in modern democratic societies, as pointed out by March and Olsen and others.

In the institutional perspective, on the other hand, as described by March and Olsen (1995, p. 30), "the axiomatics for political action begin not with subjective consequences and preferences but with rules, identities, and roles; and a theory that treats intentional, calculative action as the basis for understanding human behavior is incomplete if it does not attend to the ways in which identities and institutions are constituted, and interpreted."

The institutional perspective emphasizes the idea of "shared meaning" in two senses. In one sense we may talk about shared meaning in terms of values and world views which are related to a more or less homogeneous culture or life world. In another sense, shared meaning may refer to a common understanding of institutions and procedural rules, an understanding which does not necessarily require shared values in the sense first-mentioned.

This distinction between two forms of shared meaning is important, and is strongly related to the paradigm of communicative rationality and the notion of rationalized life worlds suggested by Habermas (1981). The term 'communicative rationality' conveys a notion of reason as intersubjectively constituted. Mutual understanding is the overarching norm, in contradistinction to the means-end efficiency criterion of instrumental rationality. The possibilities of action coordination, conflict resolution, and the mediation of collective and individual interests in modern pluralistic societies rest on the existence of a rationalized life-worldly perspective in which procedural agreement is possible without there necessarily being a substantive value-based agreement between people. As argued by Habermas, the life world components, culture, society, and personality play an important role in maintaining meaning, social solidarity, ego strength, and individual/civic action competence.

Furthermore, the precondition for complex regulation based on systemic integrative mechanisms – economic incentives and administrative power – is the existence of a rationalized life world, i.e., one in which individual as well as collective learning processes take place through communicative action. According to this view, the creation of understanding, the co-ordination of action, and the process of socialization are based on speech or language. Language therefore becomes increasingly important as the medium through which cultural reproduction (transfer and renewal of cultural knowledge), social integration (formation of social norms and legitimate order), and socialization (formation of personal identity) are implemented.

TWO VERSIONS OF INSTITUTIONALISM

When focusing on rationality, one should make a careful distinction between two versions of institutionalism. The first kind of institutionalism was developed in the wake of the concept of bounded rationality as stated by Simon (1956, 1972, 1983). That is, the core idea of bounded rationality is related to the psychological and empirical insight that people cannot make optimal decisions because they lack information and calculating capacity. Rationality is bounded negatively due to individual cognitive deficiencies. Institutionalism takes over – metaphorically speaking – the void created by the theory of bounded rationality, filling it with institutions, thereby embedding rationality in an institutional context, but without questioning the core idea of rationality itself. Instead of a negative restriction, called lack of wit, we now have the positive context of institutions. But here rationality is still understood as that of individual maximizing behaviour. The institutional context is not thought to contribute anything to rationality; rather, it is a restriction. The embedding of rationality in an institutional context, based on the notion that it helps predict behaviour, is in principle no different from adding social norms to attitude models. The justification of the model is coupled with its quality as a means of behavioural prediction and social intervention.

The second and more radical version of institutionalism goes a step further in seeing institutions as the major explanatory factor for human action. According to this view, individual action is determined by rules, norms, habits, roles, identities, etc. Rather than calculating consequences, people are believed to act on notions of appropriateness, as argued by March and Olsen.

One of the defining characteristics of this version of institutionalism seems to be that appropriateness, as the matching of situations and identities, forgoes the concept of rationality. It does so, however, in an ambiguous manner. First of all, March and Olsen are talking about two logics: the logic of consequentialism and the logic of appropriateness, respectively, thereby alluding to a kind of rationality in both, while at the same time applying the *term* "rational" only to the logic of consequentialism. But more than that, the logic of appropriateness also covers an extremely broad spectrum of human action: from nearly thoughtless habitual behaviour to theoretical and practical reasoning and arguing concerning norms and rules.

The dichotomy – instrumental vs. norm-guided action – is, it might be argued, at least partially misleading. Misleading to the degree that the calculation of consequences may itself be institutionalised as a norm during the course of history. One might therefore argue instead that an *alternative concept* of rationality is at stake here.

This standpoint is actually enforced by the institutional point of view's own logic. According to this perspective, instrumental reasoning and consequential logic over time becomes an institution in itself, due to processes of socialization and learning. It comes to programme people's perception of the intervention as well as the accounts they give of their own behaviour. And in time it influences value orientations, identities, and action competencies. But if this is so, then preferences and values can no longer be regarded as exogenous – as assumed by the paradigm of instrumental rationality. According to an institutional perspective, the means used to influence people's behaviour not only influence their overt behaviour, but also their identities. The instrument of intervention not only conveys a message, but is itself a message. And this is essentially what makes Rousseau's point of view more sophisticated than the Hobbesian model. Rationality is not first and foremost biologically, but culturally, determined. Man is a rational being because he is a social being – not the other way around – as pointed out by Mead (1934).

Hence, in order to discuss behavioural regulation as a tension between consequentialism and appropriateness, between ruler and ruled, or between public authorities and civic society, one needs a more comprehensive concept of rationality than that delivered by the exchange paradigm. And there is also a need to further elaborate the concept of appropriateness if it is to be regarded as an alternative view on rationality. The concept must be reconstructed in the vein of the distinction in moral theory between consequentialism and non-consequentialism. The Kantian distinction between hypothetical imperatives and categorical imperatives is certainly not a distinction between something more and something less rational.

One path for further development of the concept of appropriateness might be found within the framework of a theory of communicative action, as suggested by Habermas (1981, 1983). A rationalized life world is not necessarily one in which a homogeneous culture reaches consensus, but one in which the understanding of institutions and the consensus on procedural rules are established communicatively. This view is also mirrored in the account of shared meanings and social regulation given by March & Olsen when these authors argue that governance in an institutional perspective involves three important concerns, i.e., the development of capabilities, the development of meaningful accounts, and the development of identities (1995, pp. 28, 45-47). The development of capabilities concerns such important issues as the empowerment of citizens, the creation of conditions making it possible for individuals to take action, and the active support of citizens' organizational capacities. The development of meaningful accounts of societal problems as political issues is a major concern, because the actual construction of the political agenda also provides an important link between citizens and government. Finally, the development of identities is important, because democratic societies depend on active and informed participation by citizens and organized groups. And active and reasoned participation is only encouraged when, in addition to the criteria of effectiveness and efficiency, one takes into account the development of the identities themselves. As stated by March and Olsen (1995, p. 45-46) "Preferences, expectations, beliefs, identities, and interests are not exogenous to political history ... It is the responsibility of democratic government to create and support civic institutions and processes that facilitate the construction, maintenance, and development of democratic identities, and to detect and counteract institutions and processes that produce identities grossly inconsistent with democracy and therefore intolerable from a democratic point of view." As stated a little later, "the self is not only a premise of politics but also one of its principal creations" (p. 50).

The argument in favour of this view of governance essentially echoes

Rousseau's idea that in the cause of history, instinctual action is substituted by moral action. More specifically, the experience of democratic governance itself contributes to this transformation. And since the way in which regulatory authorities intervene in people's affairs influences roles and identities, governance must consider its own role in the process of governing.

SUMMING UP OUR POSITION

There is a need to discuss strategies of behavioural change in a reflective mood, taking into account that the instruments of change not only alter behaviour but also people and the relations among people. In order for such a discussion to become fruitful, we need a broader understanding of the concept of rationality. We need a concept capable of coping with the fact that it might be perfectly reasonable for people not to obey the proposition to maximize expected utility. Individual maximization of expected utility may be rejected for good reasons – good reasons rooted in a conception of civic identities which does not fit into the narrow utility-maximizer model. More specifically, we are pointing to the establishment of certain procedural norms of appropriateness linked to the concept of citizenship. The fundamental challenges of social regulation in a pluralistic society are the reconciliation of collective and individual values, beliefs and interests, and the successful co-ordination of human action. In order to meet these challenges, reasoning and arguing must not only be an element of the intervention strategy itself, but more than that, they must be one of its primary aims.

IMPLICATIONS FOR ENVIRONMENTAL POLICY AND ENVIRONMENTAL RESEARCH

The aim of this paper has been to point to limitations – and risks – in founding public policy in the environmental field wholly or predominantly on a model of instrumental rationality. Most attitude-formation and change models, which govern much of the research into the relationship between the consumer/citizen and the natural environment, are of this kind. Hence such thinking will also – or so we must assume – leave its mark on much advice given by researchers to public policymakers.

Of course, interventions in the form of campaigns aimed at providing citizens with concrete advice on how to protect their immediate environment – how to save water or energy, how to recycle, etc. – are legitimate and often needed. Furthermore, if one can establish that people's attitudes and values make them carry out instrumentally oriented behaviour damaging to the environment, it is also difficult to argue against persuasively oriented campaigns that attempt to direct these attitudes and values in a more environment-friendly direction.

But the position taken in this paper implies that we think it appropriate to warn against making such attempts the main element in governmental or quasi-governmental interventions. The development of competent, responsible citizens cannot be based on such interventions; there is even the risk that people become less competent and less responsible if interventions of this kind are allowed to dominate society.

The ultimate aim must be an increase in the number of communicative activities among citizens – activities with a bearing on environmental issues. Where such activities unfold, there is a much greater likelihood that "logics of appropriateness" are created: logics leading to desirable changes in behaviour. Furthermore, the social interactions among citizens, as well as their contacts – and confrontations – with politicians, authorities, and organizations, as well as with the commercial world, will make it much more probable that new definitions of the concept of "desirable behaviour" will emerge – definitions both more well-founded and more stable.

Concretely, then, our perspective leads to a plea for using relatively fewer public funds for interventions with a well-defined, specific behavioural change as their goal, and relatively more funds to support the work of voluntary groups and organizations (the "grassroots"), "green family" and "green neighbourhood" activities, etc.

This may sound like a trivial conclusion, or like knocking at an open door. After all, in Denmark (and the other Scandinavian countries) there is a long-standing tradition for public support to such activities, and much support, even today, is indeed channelled to actors, groups, and organizations in the environmental and ecological fields. What we have tried to provide here, however, are some *theoretical* arguments that explain why norm formation ought to be founded on people's own reasoning and arguing, and how such reasoning and such arguing are best fostered in self-determined social interaction.

Our main message, though, goes to the social scientists involved in environmental research. Too *much* work – including much of our own –

is, we believe, devoted to the refinement of attitude models dominated by instrumental rationality thinking, and too *little* work to investigations of how strategies can be formed that encourage personal responsibility, participatory decision-making, and the establishment of civic initiatives, and to a scrutiny of the conditions that must prevail for such strategies to be successful. Not an easy task, to be sure – many elements in our increasingly individualized societies undoubtedly work against civic involvement in matters of joint concern – but there are nevertheless success stories (as well as failures) to study and learn from. Is it not, for example, high time that Elinor Ostrom's "Governing the commons: The evolution of institutions for collective action" (1990) becomes a more often-quoted source in research on citizens' environmental behaviour than Icek Ajzen's "The theory of planned behaviour" (1991)?

REFERENCES

Aarts, H., Verplanken, B., & van Knippenberg, A. (1998). Predicting behavior from actions in the past: Repeated decision making as a matter of habit? *Journal of Applied Social Psychology*, *28*, 1355-1374.

Ajzen, I. (1985). From intention to actions: A theory of planned behavior. In: J. Kuhl & J. Beckmann (Eds.), *Action-control: From cognition to behavior*, pp. 11-39. New York: Springer.

Ajzen, I. (1988). *Attitudes, personality, and behavior*. Chicago: Dorsey.

Ajzen, I. (1991). The theory of planned behavior. *Organizational Behavior and Decision Processes*, *50*, 179-211.

Ajzen, I., & Fishbein, M. (1980). *Understanding attitudes and predicting social behavior*. Englewood Cliffs, NJ: Prentice-Hall.

Beck, U. (1994). The reinvention of politics: Towards a theory of reflexive modernization. In: U. Beck, A. Giddens, & S. Lash (Eds.), *Reflexive modernization*, pp. 1 – 55. Cornwall: Polity Press.

Bellah, R. N., Madsen, R., Sullivan, W.M., Swidler, A., & Tipton, S. M. (1985). *Habits of the heart: Individualism and commitment in American life*. New York: Perennial Library.

Brehm, J. W. (1966). *A theory of psychological reactance*. San Diego, CA: Academic Press.

Conner, M., & Armitage, C. J. (1998). Extending the theory of planned behavior: A review and areas for further research. *Journal of Applied Social Psychology*, *28*, 1419-1464.

Eagly, A. H., & Chaiken, S. (1993). *The psychology of attitudes.* Fort Worth, TX: Harcourt Brace Jovanovich.

Fishbein, M. & Ajzen, I. (1975). *Belief, attitude, intention, and behavior: An introduction to theory and research.* Reading, MA: Addison-Wesley.

Foucault, M. (1987). Questions of method: An interview with Michel Foucault. In: K. Baynes, J. Bohman, & T. McCarthy (Eds.), *After philosophy – End or transformation?*, pp. 100-117. Cambridge, MA: MIT Press.

Grunert, S. C., & Juhl, H. J. (1995). Values, environmental attitudes, and buying of organic foods. *Journal of Economic Psychology, 16*, 39-62.

Habermas, J. (1981). *Theorie des kommunikativen Handelns.* Frankfurt a. M.: Suhrkamp.

Habermas, J. (1983). *Moralbewusstsein und kommunikatives Handeln.* Frankfurt a. M.: Suhrkamp.

Habermas, J. (1992). *Faktizität und Geltung.* Frankfurt a. M.: Suhrkamp.

Hobbes, T. (1968). *Leviathan.* Harmondsworth: Penguin Books. Original work published 1651.

Lakatos, I. (1970). Falsification and the methodology of scientific research programmes. In: I. Lakatos & A. Musgrave (Eds.), *Criticism and the growth of knowledge*, pp. 91-195. Cambridge: Cambridge University Press.

March, J. G., & Olsen, J. P. (1989). *Rediscovering institutions.* New York: The Free Press.

March, J. G., & Olsen, J. P. (1995). *Democratic governance.* New York: The Free Press.

Mead, G. H. (1934). *Mind, self, and society.* Chicago: University of Chicago Press.

O'Keefe, D. J. (1990). *Persuasion – Theory and research.* Newbury Park: Sage.

Ostrom, E. (1990). Governing the commons: The evolution of institutions for collective action. Cambridge, England: Cambridge University Press.

Parsons, T. (1968). *The structure of social action.* New York: The Free Press.

Pieters, R. G. M. (1991). Changing garbage disposal patterns of consumers: Motivation, ability, and performance. *Journal of Public Policy & Marketing, 10*, 59-76.

Putnam, H. (1987). Why reason can t be naturalized. In: K. Baynes, J. Bohman, & T. McCarthy (Eds.), *After philosophy – End or transformation?*, pp. 222-244. Cambridge, MA: MIT Press.

Rousseau, J.-J. (1987). Samfundspagten. Translated from Du Contrat Social. Copenhagen: Rhodos. Original work published 1762.

Sagoff, M. (1988). The economy of the earth. Cambridge: Cambridge University Press.

Simon, H. A. (1956). Rational choice and the structure of the environment. Psychological Review, 63, 129-138.

Simon, H. A. (1972). Theories of bounded rationality. In: C. B. McQuire & R. Radner (Eds.), Decision and organization, pp. 161-176. Amsterdam: North-Holland.

Simon, H. A. (1983). *Reason in human affairs*. Oxford: Basil Blackwell.

Szerszynski, B. (1996). On knowing what to do: Environmentalism and the modern problematic. In: S. Lash, B. Szerszynski, & B. Wynne (Eds.*), Risk, environment and modernity*, pp. 104-137. London: Sage.

Thøgersen, J. (1994). A model of recycling behaviour. *International Journal of Research in Marketing, 11*, 145-163.

Tocqueville, A. (1978*). Lighed & Frihed.* Translated from *La démocratie en Amérique*. Faaborg: Forlaget i Haarby. Original work published in 1835.

Triandis, H. C. (1977). Interpersonal behavior. Monterey: Brooks/Cole.

Østerberg, D. (1980). *Samfundsteori og nytteteori*. Oslo: Universitetsforlaget.

CHAPTER 6

Rationality, Environmental Law, and Biodiversity

Helle Tegner Anker and Ellen Margrethe Basse

RATIONALITY IN LAW AND JURISPRUDENCE

Introduction

THE CONCEPT OF rationality has a number of different meanings, depending on the context in which it is to be described or analysed. This chapter will explore the concept of rationality in relation to issues of environmental law, seeking a "background" explanation and understanding of this particular area of legislation. The analysis is not based on theoretical discussions of different concepts of rationality, but rather on more pragmatic observations and assumptions about the purpose and operation of environmental law. A central element behind this is the clarification of the kind of rationality to which the legal system appeals – in other words, the rationality of law (norms) as a basis for decision-making. In this respect, law-making (including reference to goals and interests), legal and institutional structures at national, regional and global levels, and the choice of legal and regulatory measures become essential elements. One could say that the *"legal thinking"* embedded in law is a core element in any attempt to analyse rationality in relation to environmental law. But, how is legal thinking defined – if it can be defined? As expressed by Merryman, Clarke & Haley (1994, pp. 3-4), *"The legal thinking depends on the legal tradition – a set of deeply rooted, historically conditioned attitudes about the nature of law, the role of law in the society, the proper organization and operation of legal systems, and the way law is (or should be) made, applied etc. Such traditions and thinkings relate to the culture of which it is a partial expression."*

A traditional concept of rationality is often described as a behavioural pattern based on a systematic awareness of one's own preferences and of a systematic procedure for the maximum fulfilment of those preferences (cf. *behavioural rationality*). Rationality in this sense is, as a basic element of economic theory, very closely related to the preferences of individual persons acting more or less independently. Although the economic rationality concept normally has a dominating position, it is not necessarily adequate in a broader perspective, or it needs certain adjustments to fit into other contexts than those of traditional economic thinking.

From a legal perspective, e.g., the legislative system, rational behaviour is likely to be influenced by the purpose of law and the legal requirements for fulfilling it. The basic idea of the legal system can be described as a consistent and coherent body of norms whose observance secures certain valued goals which can intelligibly be pursued collectively (MacCormick, 1978). It is, however, as pointed out by Graver (1989), important to realise that other than legal factors may influence decisions made according to the law and therefore it is also important to maintain a distinction between the *rationality of norms* (law) and *rationality of actions* (decisions). The rationality of norms is a crucial concept when analysing the legal framework for decision-making, although it cannot necessarily predict the rationality of decisions.

The behaviour of the *legal agent* is interesting, from the perspective of both the individual and the system. A legal agent is, however, not easily definable. One definition could be a person (or an authority) acting in accordance with the judicial or administrative decision-making function. This (traditional) definition, however, focuses on the official application of law in the form of individual decisions, and excludes to a certain degree the addressees of general rules. But the public could also be said to act according to legal principles and rules. The lawmaker – the legislature – is also excluded from the traditional definition. However, a parliament acts within the legal framework of international obligations, the Treaty of Rome, the constitution, and general principles of law, and is thereby acting as a legal agent. And what about the rationality of legal scholars? Jurisprudence influences both lawmaking and the application of law, and vice versa – although the degree of influence may be disputed. The legal system or the living law (Sacco, 1991) thus contains many different elements and actors.

Most legal agents will not act as individuals, but as representatives of an authority, a specific branch, interest group, or the public. This fact does not exclude rational decision-making, although one could speak of collective rationality. The process of balancing different interests, for example, will often involve rational decision-making behaviour in order to determine which interests are relevant, and how they are to be balanced. A goal-orientated cost-benefit analysis could be very useful in such a situation – although not always satisfying. The institutionalisation of either collective or individual interests, values, or preferences does not necessarily change the idea of rational behaviour. Sagoff refers to the statement, "*... that the promotion of (particular) private interests is often a legitimate function of regulation, but it is to say that the administrator must deliberate about those interests, rather than responding mechanically to constituent pressures*" (Sagoff, 1988, p. 11).

In relation to law – and environmental law in particular – there seem to be two different categories of *rationality*. The *first category,* known from the works of Weber, relates to the formal method of following criteria of success laid down in advance by the legislator – expressed by *norm-orientated rationality* and a subsumption of logical methodology (Ross, 1953). Such a rationality concept is common in legal theory, in relation to certain kinds of legal positivism. This category may also reflect a *goal-orientated rationality,* which in many ways can be said to reflect the traditional economic concept of rationality – but it is not necessarily based on individual preferences. In fact, the question can be raised of whether rationality is not always partly goal-orientated, and whether certain elements of both norm- and goal-orientated rationality will not always be prevalent in a legal context. Weber has, however, also written of a value-orientated rationality, in relation to which the decision-making, or way of acting, in itself represents certain values. According to this perception, there is a *second category* of rationality that may relate to deliberation, fairness, equity and virtues of clarity and open-mindedness in describing and finding methods of problem-solving (Sagoff, 1988, p. 13 and p. 222). In some ways, this last category corresponds to the concept of *communicative rationality* introduced by Habermas, as well as to Teubner's theory of *reflexive law.* Teubner's theory has been said to imply a concept of law that gives law as such a life and consciousness of its own, and in which different types of law (or regulation styles) constitute different communication structures with their own specific rationalities, so that the actors communicating are tied up to the different structures (Graver, 1989, p. 79). The second category of rationality includes certain moral aspects of the relationship to other actors, to future generations and to nature. This category may also be characterised by a larger number of procedural aspects in order to create a proper framework for communication and deliberation. Yet there are no clear distinctions between the different types of rationality.

Environmental law seems to be an example of the array of problems and interactions relevant to the concept of rationality. In the following, we will try to provide an insight into some of the most prevalent elements of law and jurisprudence central to the evolution of environmental law and to the concepts of rights and interests. Finally, the legal implications of the concept of biodiversity will be analysed as an example of the complexities in environmental law, and perhaps as indicative of new directions in legal rationality. The aim of this article is not to give a clear, precise answer as to which type(s) of rationality are, or should be, prevalent in environmental law. Rather, the aim is to provide some reflections on how (rational) decision-making is, or can be, influenced by a number of different factors in the legal system.

Rationality and environmental law

Law as such has been identified as the enterprise of subjecting human conduct to the governance of rules (Fuller, 1964, p. 106). The purpose of law, however, may vary from time to time, since law is not static (Sacco, 1991, p. 390). The distinction between the general purpose of achieving a stable society and the more specific purposes of different types of legislation, e.g., the purpose of environmental law, is very important. Certain basic objectives or elements of law are reflected in constitutional principles, e.g., the protection of private property rights, or in other fundamental principles of law, e.g., the principles of legality, proportionality and equality. Western European law is generally characterised by a liberal attitude, which means that due account should be taken of private economic interests which are particularly affected by the rules of law or the specific decision. It should therefore be stressed that law and the legal system serve to protect the existing (economic) rights and interests *of the individual person*, e.g., rights of freedom and rights of private property, as well as the rights associated with the concept of legality. The rationality accepted in such systems is a behavioural rationality closely related to the preferences of individual persons. As an example, damage to the environment, which does not in itself have an economic value, but may have great value in other terms (such as the loss of an – in economic terms – insignificant species), cannot be compensated for if the law (as is the case in most European countries) only allows for compensation in terms of economic loss.

The general purpose of environmental law will often be to protect nature and the environment from harmful human activities – although not all environmental degradation can be ascribed to such activities. But to what extent this protection reflects only the (economic) interests of human beings, e.g., the need for clean air and water, or also reflects the interests of nature as such, may be debatable. That is, the discussion centres on an anthropocentric, versus an ecocentric, approach. This distinction will, however, not be further elaborated on in this chapter.

The evolution of environmental law shows an expansion of the various interests to be integrated into decision-making at all levels – from the conflicting (economic) interests of neighbours, via the interests of public health, towards the interests of present and future generations (intra- and intergenerational equity) and of nature as such (intrinsic value of biological resources). The number and character of the different interests involved in (environmental) decision-making vary in different societal and legal contexts and require consideration of how the representation and balancing of different interests is to be carried out. The law, reflecting the core values of society, creates a framework for such decision-making, and for the rationality of decision-makers (rationality of norms). This includes explicit or implicit reference to environmental principles and objectives, to substantive and procedural requirements, to the choice of regulatory instruments, and to the organisational structure of the administrative and judicial system.

Yet, the substantive elements of a legal system are seldom unambiguous – they are not always clearly expressed, and they often represent conflicting objectives. One of the main concerns of environmental law is finding a balance between the need for development and the problems of environmental degradation (Preston, 1995, p. 228). The intervening norms in environmental law call for a balancing or reconciliation of such contradictory interests. Furthermore, the introduction of the concept and objective of sustainable development adds to the complexity, since the concept in itself requires a balancing of different, contradictory aims, as is also the situation in relation to the concept of biodiversity. The complexities may be further accentuated when new rules are introduced into existing legislation or into legal systems directed towards solving other problems than those giving rise to the new rule. Borrowing principles and institutions from more traditional parts of legal systems may give rise to strange and artificial rationalisations (Sacco, 1991, p. 399).

The different objectives or purposes of legislation may not only be referred to in the general substantive requirements of law, they may also be expressed by the choice of legal and regulatory instruments in a specific area of regulation. The complexity of environmental issues tends to reduce our ability to define a "correct" regulation on an *a priori* basis. The outcome of most environmental conflicts cannot be deduced from the norms or goals laid down, as the "correct" solution will depend on an individual balancing of different interests. On the basis of this assumption, there is a need to ensure the best possible basis for such an interest-balancing process. This means that procedural requirements, such as participation, information, and assessment procedures, become an important part of both the legislative and administrative process – often emphasising a communicative rationality. It is, however, important to realise that there are different trends and rationalities used in environmental law – and they appear in varying strengths in different regulatory systems.

The organisational structures in decision-making systems may also reflect a balancing of the different objectives and interests involved. This is perhaps particularly true in legal or regulatory systems characterised by discretionary powers. As an example, two different trends in Danish public environmental law have been identified as the trend toward administrative co-ordination and the trend toward administrative corporatism or co-operation, each characterised by the establishment and organisation of administrative systems involving a number of different actors (Basse, 1987).

It is important to point out that in the environmental law-making of the Scandinavian countries, with their long tradition of collective bargaining, participation by the private parties concerned is common. Some of the rules state which legitimate interests are to be included in the decision-making process, e.g., via the organisation of the decision-makers as well as the process organisation. Other rules may indicate that the relevance of certain interests should be limited and scaled accordingly (Basse, 1987). In this respect, the approach is based on a communicative rationality focusing on the different means of interest balancing. For the purpose of this presentation, the most important point is that the conflict of interest is not resolved solely at the level of the legal system – its boundaries are transcended, and resolution of the conflict is decided by factors other than law alone.

Rationality in law-making

The traditional doctrine in Western European countries (the legal dogmatics) relies largely on a strictly defined separation of powers between legislative, executive and judicial institutions. In that framework, the task of government becomes very clear: once democratically established, the legislature has the responsibility of formulating laws, which the executive institution then implements, and the judiciary system ensures that they are consistently applied.

In principle, legislation in Denmark is based on the concepts of representative democracy. As such, statutory law is the result of trade-offs between legislators representing various interests. Law-making, however, is also undertaken within the framework of constitutional and international principles and obligations. Furthermore, as has been argued by Fuller (1964), law-making is subject to a number of formal moral requirements – the "inner morality" of law. Whether law-making is also subject to more substantive moral requirements has likewise been the subject of ongoing discussions between the theoreticians of legal positivism (e.g., Hart & Ross) and of natural law (e.g., Finnis, Beyleveld & Brownsword). It seems clear that (environmental) legislation today reflects moral requirements through the concept of sustainable development and the principles of intra- and intergenerational equity – both as a response to the international framework, and to the societal concern of sustainability issues. The manner in which this is to be reflected, and whether this implies reconsideration of the legal system as such, remains an open question.

The very nature of environmental problems makes it most difficult to enact precise and efficient legislation. Furthermore, the number of different – strong – interests involved in environmental law-making, both at national and international level, makes compromise and framework-legislation very common phenomena. A government is often interested in gaining more practical experience in order to establish a framework of parliamentary decisions, action plans, strategies, etc. Such types of vague, imprecise environmental rules do not necessarily respond to the above-mentioned inner morality of law. Important causes of insufficient environmental regulation may also be identified as: vague, ambiguous and conflicting goals in legislation; lack of knowledge about complex cause-and-effect relationships; and decision-making based on insufficient or misleading information.

In some European countries – e.g., the Netherlands, Sweden and

Denmark – consultations with branch- and interest-organisations are a normal part of the law-making procedure. Such consultations, however, do not necessarily lead to an open law-making process, and may fall short of including weaker or less obvious interests.

The complexities of environmental issues, and the requirements of responding to the concept of sustainable development, make law-making a very complex process, possibly requiring a rationality capable of securing the involvement of a number of different and contradictory interests and objectives.

Rationality in jurisprudential analysis

It is also relevant to discuss the concept of rationality in jurisprudential analysis or legal science, which is normally characterised by a strong interaction with law, e.g., by clarifying the meaning of law, or by pointing out needs for legislative initiatives. One can distinguish between the components of jurisprudential analysis which mainly register certain (legal) facts and the components that seek to achieve certain objectives. The former category is based on an assumption that the sources of law can be seen as naturally occurring phenomena, or data, in the study of which the scientist can discover inherent principles, concepts and relationships (legal positivism). The latter category is based on an analysis of the role and functions of law in a broader perspective. Although there is no clear dividing line between the two categories, it appears that moving from the first category to the second gradually extends the scope of analysis and increases the relevance of different concepts of rationality. MacCormick (1978) has identified three central elements of legal reasoning; namely deductive justification, arguments of consistency, and arguments of coherence, in relation to which the degree of rationality or reasoning increases. One could view jurisprudential analysis from a similar scale of rationality – perhaps also indicating different types of rationality.

Environmental law can, like other areas of law, be analysed from different jurisprudential angles. It is possible to make a (narrow) dogmatic or positivist analysis of the interpretation and application of certain areas of environmental law. Such an approach is, however, likely to fall short of discussing central elements of environmental policy, such as sustainable development and biodiversity. It is also possible, and perhaps more often happens, that a broader view on environmental law is taken, and factors other than law alone, e.g., policy statements and ecological understanding are considered as a means of achieving a functional environmental law. Finally, it is possible to let an idea of ecological modernisation penetrate into the jurisprudential analysis, and to focus on the need to restructure the legal system by changing its core values. Depending on the premises, it might be appropriate to speak of different types of rationality, characterised by different sets of premises. In this respect, both the awareness of preferences and the use of a systematic procedure are important. In practice, however, jurisprudential analysis of environmental law will often be based on a mix of different approaches.

The need for a new orientation in juridical research – e.g., in research on environmental law – is primarily based on changes in modern society and in the legal systems, requiring a more general view of the functions of law. Compared to traditional jurisprudential thinking and terminology, environmental law – where the protection of common resources becomes essential – may use different concepts – e.g., in relation to the conventional sovereignty of states, property rights, and responsibility. The right of state and landowner, respectively, to use natural resources is delimited, and the obligation on the part of both state and citizens to care for common natural resources is added. Environmental problems also require the use of a long-term perspective and holistic approach.

ENVIRONMENTAL LAW AND THE CONCEPTS OF RIGHTS AND INTERESTS

Introduction

The concepts of rights and interests are central elements of law and of the understanding of the rationality expressed in environmental law. They are, however, not sharply defined. Traditionally, the term interest is interpreted in a legal context as that of a specific claim, and as such, very strongly related to the concept of rights. The following discussion takes the traditional legal perspective of *rights* as representing a concept from which one can derive a duty to *respect,* or not to violate, the specified rights, with a typical example being the constitutional protection of private property rights. Yet, one will discover that such rights are normally not static. *Interests* are, however, considered as a wider concept, from which one can derive a duty to take into *consideration* specified interests. It appears that changes in the position of rights and interests are reflected in the evolution of environmental law, e.g., by an increase in the protection of interests and by a more complex interplay between such interests.

The environment can be the object of legal regulation, e.g., in terms of the *right* to own and use natural resources as well as in terms of a *duty* to protect the environment. But only humans can be holders of rights in most European legislative systems. All other beings are normally ascribed rights or values indirectly, relative to man. The question of directly ascribing rights to natural assets, e.g., endangered species, raises a number of fundamental questions in most legal systems. Although a species may have a right to survive, this does not necessarily entail a human obligation to act for its survival – but it might entail a duty to refrain from damaging activities. In this way, the survival of a species becomes an *interest,* which must be taken into account.

The traditional focus on the economic rights and interests of individuals is not easily combined with the new elements in environmental law, e.g., in relation to sustainable development and biodiversity, where the interests of future generations and of nature as such are central elements. Thus, as indicated above, the concept of interest in environmental law may have a different meaning than in the traditional legal concept. In this respect, interests falling outside both individual economic interests and the traditional legal sphere need protection by the legal system, e.g., by substantive obligations to take such interests into consideration, or by procedural requirements ensuring the possibility of reflecting such interests in the decision-making process. Such protection of interests may require a rationality-based communication or deliberation, since the interests of future generations and or of nature as such can hardly be described in an objective way, or be subject to traditional cost-benefit analyses. In this perspective, the concept of interests refers more to the duty or the responsibility of taking such interests into consideration – rather than to their legal rights (with regard to protection).

The question still remains of how interests that are not directly related to individuals or the immediate sphere (environment) of the individual person are to be reflected in law and decision-making at all levels. Is it at all possible to determine the interests of future generations, or of nature as such, and if so, who is capable of doing so? The uncertainty attached to non-economic interests or values will, in traditional economic terms, lead to a discounting of such interests in promoting present (economic) interests (Norton, 1998).

The legal protection of interests can be divided into *direct* protection (e.g., substantive requirements), and more *indirect* protection, secured by requirements allowing the different various interests to influence decision-making (e.g., by means of principles, guidelines, assessment procedures, access to justice and other procedural requirements). In order to protect the largest number and variety of interests, the legal framework must encourage open, informed decision-making (communicative rationality). The protection of interests naturally depends on the character of the different stakeholders, e.g., organisations, citizens, authorities, future generations, and nature as such. The legal protection of vague, imprecise interests, such as those of future generations and nature, should be characterized by a rationality expressing the virtues of reflection, consideration and responsibility.

The evolution of environmental law – Danish experience

The evolution of environmental law emerged from the neighbour nuisance law, through public laws on planning and environmental protection, supplemented by EU competition and environmental law, and finally moving towards a more holistic and global perspective, e.g., in relation to sustainability and biodiversity. The environmental law of today will reflect these different types to a varying degree.

The different types of environmental law are related to perceptions of societal concerns triggered by developments in relation to *inter alia* technology, information and internationalisation. The scope of knowledge has increased drastically, as have the complexities of environmental problems and their possible solutions, and this is reflected in the development of environmental law. Different phases of environmental law may, *inter alia,* by their perception of rights and interests be related to different types of rationality. In many cases there has been a tendency to adapt new demands for regulation to existing regulatory systems, and conflicts may arise from such an approach, e.g., between traditional principles of law, and new environmental principles incorporated into existing regulatory systems. Such conflicts can perhaps be explained by the differences in the rationale or rationality behind various principles and systems.

Neighbour- and environmental liability law

Environmental law has emerged from the traditions of solving conflicts between neighbours and property owners arising from nuisance caused by one property owner to another. Neighbour or nuisance law is traditionally defined as the sum of legal rules limiting the owner's rights to use his own property and expressing his duty to respect the rights and interests of adjoining property owners. Its purpose is to protect the economic values or interests of neighbouring property, rather than to protect public interests or societal values, although in some cases there may be overlapping interests. Activities which constitute an annoyance or disturb one's use of the property or substantially make its ordinary use uncomfortable may be nuisances, and one owner may bring an action against his neighbour within the limits of these liability rules. Thus the plaintiff must demonstrate that he has suffered economic loss as a consequence of a cognisable damage. Such claims generally seek prohibition of the emissions in question if they are not in compliance with legal requirements or the ordinary level of environmental quality in that area. Demands for pollution reduction or remedial measures may be imposed in order to bring the affected property back to its original condition, and payment of financial compensation may be imposed. Such a judgement involves determining whether the interference is *unreasonable* in the sense that the harm to the plaintiff is greater than the utility of the defendants' conduct. The general rule is that the owner of damaged property can demand compensation for restoration only if the cost of restoration is not out of proportion to the value of the damaged property.

A similar line of thought characterises environmental liability law, which is focused on the economic rights and interests of individuals – not only of neighbours. Environmental liability is in many countries based on general liability rules, i.e., a broad standard of fault and negligence. Strict liability is the exception in civil law as in common law systems (Sacco, 1991, p. 358). Negligence (culpability) exists when the person causing the loss has not exercised the caution required of "the rational man". A finding of fault depends on whether the party has a duty to behave according to a certain standard of care. Fault may consist of either an action or a failure to act when action is required by environmental law (including failure to obtain a licence prescribed by law). Therefore, in the field of environmental liability, there is a strong interaction between fault-based liability and public environmental regulations. Property owners must take into consideration the sensitivity of the area, and maintain buildings and technical installations in good repair in order to prevent unacceptable nuisance.

Generally, the burden of proof lies with the injured party, i.e. he loses his case if he cannot prove negligence on the part of the polluter. In connection with environmental damage caused by particularly risky activities, the legal precedent is thus characterised by very strict assessment of cautionary measures and the reversal of the burden of proof. Negligence in itself is not sufficient to create liability. Liability requires causal connection, proportionality/adequacy between action/inaction and the actual pollution/damage as well as proof of financial loss.

A specific aspect of the adequacy requirement is the fact that compensation can normally only be *claimed by individuals* who have an interest in the polluted/affected property or the values connected to the property. Under traditional liability law, public authorities cannot claim compensation for general loss of existing flora and fauna, as such goods are not the property of the public authorities.

Public legislation on the environment and protection of nature

Present-day environmental law consists primarily of legislation on planning, nature protection and environmental protection. The purpose of environmental protection has primarily been to protect human health interests by regulating polluting activities and substances, *inter alia* to avoid or reduce pollution of water, air and soil. Legislation on nature conservation or nature protection has primarily had the purpose of conserving certain landscapes and natural elements, while planning legislation has a co-ordinating role. The main thrust of these types of regulation has been a massive body of public laws safeguarding public health and recreational interests. Public authorities have the power to impose duties or restrictions on individuals in order to protect such interests, e.g., via licences, emission standards, conservation orders etc. The scope of environmental and nature protection law has gradually been expanded so as to cope with areas even beyond immediate health and recreational interests, e.g., by the introduction of concepts of sustainability and biodiversity.

Generally, the rationale behind this type of regulation is characterised by an interest-based approach. Procedures and rules governing the organisational structure play a central role in relation to interest-balancing, as substantive rules are few and vague. Environmental statutes normally comprise an introduction to the main objectives and (environmental) principles for the actual regulation (goal-orientated norms), and for the balancing of interest in general as well as specific situations (balancing norms). A number of environmental principles are stated in the opening provisions of some environmental statutes. Among these is the precautionary principle, the 'polluter- pays' principle, the principle of tackling environmental problems at the source and the principle of cleanest or best available technology. The implementation of such principles has, however, not been especially significant, as their impact is limited in comparison with the principles of traditional unwritten administrative law, which is primarily based on individual law and order considerations (Basse, 1996).

Thus, conflicts between the interest-based approach and a more traditional rights-based approach have occurred in some cases. One example is the polluter-pays principle – based on the economic rationale of internalising environmental costs. In Denmark, a number of cases show that this principle is tied into more traditional fault-based liability considerations – focusing on the protection of the rights of "innocent" individual persons. This has now been clarified in a new act on soil pollution.

The concept of sustainable development

The concept of sustainable development was introduced by the Brundtland Commission in 1987. It has been further elaborated in several international settings, including the 1992 Rio Declaration, and it is also reflected in national legislation in several countries, e.g., the introductory statements of Danish environmental statutes. However, Denmark has had only limited experience with the implementation of the objectives of sustainable development.

Substantively, the concept of sustainability implies a call for the inclusion and reasonable prioritisation of more "soft" interests in the balance. This balance will influence the assessment procedure as well as the interpretation of legislation at the national level. As an example of sustainability, the consequences of the traditional definition of the terms "rights", "interests", "responsibility", and "liability" may be stressed, through a redefinition of legal concepts on the basis of demands for sustainability. When utilising a sustainable development or sustainable growth concept, *the balance of interests* is dependent on the interplay between considerations of *inter alia* equality, jurisdiction, trade, development and environment.

Procedurally, the decision-making situation is, in accordance with the sustainability concept, assumed to be changing, i.e., in a direction ensuring a more open decision-making system and better informed decision-makers and citizens, as expressed in the 17th principle of the Rio Declaration. It has been argued that the key issue in the debate on environmental principles is that sustainable development is a permanent process, depending on the measures taken, and therefore requires an emphasis on process and instrumentality issues (Mann, 1995). Public participation and access to justice – for individuals as well as for environmental organisations – in modern environmental law must accordingly be considered as one of the vital conditions in fulfilling demands for sustainability.

Legal requirements for a good environment imply, furthermore, that citizens have a right to information on environmental issues of importance to their health and welfare. Today, such legal claims are formalised at international, Community and national levels – especially through the 1998 Aarhus Convention. Strategic environmental assessment and environmental impact assessments can be used as means of securing integrated preliminary assessments at policy, programme, plan, and project levels. International standards for such assessments have been motivated by demands for sustainability, e.g., the Espoo Convention, and the 10th principle of the Rio Declaration.

Administratively, the sustainability concept, as indicated, includes a change in the existing sectoral limits between different authorities and other actors – including better co-ordination, increased co-operation, and external integration of environmental principles, considerations, and management in all private and public decision-making systems. One could therefore argue that the holistic interest-balancing character appeals to a communicative rationality.

Reflecting the variety of legal systems, incorporation of the concept of sustainability will be carried out in accordance with *the legal tradition of each state*. The implementation of a sustainability objective by the use of environmental principles in national legislation will be greatly dependent on systems of decision-making, and on the measures which national authorities must use in realising the intentions. Contrary to the concept of law and order, the sustainability and biodiversity concepts are not primarily based on a need for individual protection of rights against a strong public authority, but rather on a need for protection of life-sustaining systems for present and future generations, the availability of resources for human well-being, and their rate of consumption. In practice there are numerous legal barriers based on traditional law and order principles and traditional "legal thinking" or rationality, which are very important when such international obligations are brought into national law.

BIODIVERSITY, LAW AND RATIONALITY – AN EXAMPLE OF THE COMPLEXITIES OF ENVIRONMENTAL LAW

Introduction

The concept of biodiversity has evolved in the international arena in recent years – leading to the signing of the Convention on Biological Diversity at the Rio Conference in 1992 (hereafter the Biodiversity Convention). The objectives of the Convention are not only conservation and sustainable use of biological resources, but also fair and equitable distribution of the benefits arising from the utilisation of genetic resources (art. 1). The concept of biodiversity is far-reaching: the diversity of ecosystems, the diversity of species, and the genetic diversity within species. Although it might seem possible to make lists or inventories of the biological resources – if only we had sufficient knowledge – the concept remains very complex. For instance, is it possible to protect one species without causing adverse effects to other species or to genetic diversity? And how do we deal with conflicts between species, populations or ecosystems?

Apart from being a complex concept in itself, the objectives stated in the Biodiversity Convention represent different approaches to biodiversity protection. The overall objective is, of course, to protect or conserve biological diversity in general. Conservation may in this respect be defined in both a broad and a narrow sense. The broad definition, according to the 1980 IUCN World Conservation Strategy, embraces issues of sustainable use, while conservation in a narrow sense is confined to more traditional nature protection or conservation activities. A key concern of biodiversity protection appears to be to avoid, minimise or reverse the adverse effects of human activity. In this respect, sustainable use of biological resources becomes central to biodiversity protection. Thus, efforts to secure sustainable use supplement more traditional measures of nature conservation. The legal definition of sustainable use is laid down in Article 2 of the Convention as

> "… the use of components of biological diversity in a way and at a rate that does not lead to the long-term decline of biological diversity, thereby maintaining its potential to meet the needs and aspirations of present and future generations".

Efforts to protect biodiversity (or portions of it) have a long history – both in international and national law (Bowman, 1996, p. 7). Historically, these mostly relate to the protection (and management) of fish and game – that is, biological resources directly exploited by traditional human activities – fishing and hunting. The concept of biodiversity, however, represents a new line of thought in relation to the management of the biological resources of the Earth. This new line of thought is closely related to the concept of sustainable development, which requires a holistic approach, integrating economic, social and environmental considerations and respecting the needs of future generations. This inter-relationship is apparent in relation to the requirement for sustainable use of biological resources and to the principles of intra- and intergenerational equity, as reflected in the Biodiversity Convention.

Degradation of biological diversity is seldom immediately recognisable, and is often the result of diffuse and cumulative activities (Snape, 1996, p. 38). Environmental law, on the other hand, has traditionally evolved as a response to more immediate environmental problems affecting the quality of human life, be it degradation of the quality of air, water and soil, or disturbances of the aesthetic or cultural quality of nature and the environment. The concept of biodiversity, characterised by a large degree of uncertainty, penetrates into many different sectors and requires a long-term, integrated, and holistic approach. A careful balancing of different interests and values becomes central to the conservation and sustainable use of biological resources, keeping in mind the precautionary principle as reflected in the Rio Declaration and in the preamble of the Biodiversity Convention. The "soft" interests of biodiversity cannot, however, easily be explained or balanced in rational economic terms, thus considerations in a broader – ecologically informed – context become central (Norton, 1991, pp.140-154).

The concept of biodiversity not only implies consideration of the economic (or utilitarian) interests of the present and future generations, but also the consideration of the intrinsic value of biodiversity as such – i.e., purely non-anthropocentric values. Although the Convention can be characterised as representing an anthropocentric viewpoint, consideration of the intrinsic value of biodiversity is reflected in the preamble. However, it is debatable whether this notion reflects more than an acceptance of a moral considerability (Bowman, 1996, p. 29).

The aspects of uncertainty and moral considerations as key elements of the biodiversity concept are crucial to understanding what kind of rationality must be part of normative regulation and decision-making.

Biodiversity and moral considerations

It has been argued that environmental law in general (Sagoff, 1988, p. 6), and biodiversity law in particular (Snape, 1996, p. 218), are not necessarily based on the traditional rationale of government intrusion, namely that of making markets more efficient by allocating resources. Many of the problems in effectuating environmental law may be ascribed to this focus on "the calculus of the market". However, so far, environmental law has often managed by adapting new principles and measures to existing legal systems – perhaps sometimes without realising the problems of implementation. Consequently, a legal framework may be established without proper deliberation on how and at which point to carry out the balancing of interests and values. Sagoff has argued that some statutes explicitly set environmental goals without regard to the costs, benefits, obstacles, or means involved in achieving them (Sagoff, 1988, pp. 199-205). He further stresses that ... *if we continue to cast social and environmental policy entirely in the optative mood, the perfect society to which we aspire in theory may become a powerful enemy of the good society we can become in fact* (Sagoff, 1988, p. 200).

The concept of biodiversity seems to illuminate the problems inherent in many areas of environmental law. This is partly due to the fact that this concept appears more complex than the traditional range of environmental aspects, e.g., nature protection or water, air and soil pollution. Furthermore, the biodiversity concept requires a reassessment of policies and practices in a number of different sectors or spheres, e.g., production, energy, transport, waste, agriculture, land use, conservation, development, economics and trade (Johnston, 1997). In fact, the concept and the implications thereof are not easily definable. The Biodiversity Convention now constitutes a starting point, although it, too, is the result of many trade-offs on the international political arena (Koester, 1997).

Another reason is the moral considerations attached to the concept of biodiversity. The concept of *moral considerability* has been presented as a response to philosophical discussions on the status or intrinsic value of nature (Callicott, 1986). Biodiversity protection not only refers to the immediate sphere of human activities, but also to the protection of species, genetic diversity, and ecosystem processes not yet within the reach of human knowledge. The latter does not necessarily refer to an ecocentric approach; although not having any immediate meaning for human beings, species, genes or processes as such may take on such meaning in a long term perspective, whether for future generations or as part of what has been called ecosystem resilience (Perrings et al., 1995 p. 5). The concept of moral considerability, therefore, may not only refer to moral considerability towards nature, but may also reflect the consideration of the opportunities and constraints that are left for future generations. We cannot at present determine the needs of future generations, but we can consider and enhance the range of opportunities available to them (Norton, 1997). In other words, moral considerability is not necessarily a question of ascribing (legal) *rights* to animals and other elements of nature (as advocated by Stone, 1972), but one of considering and protecting the *interests* of nature as such, and the interests of the generations to come. In this context, one may argue that the rationality of norms and decisions must be based on the duties of conscious action and deliberation, thus focusing on the ways in which individual cases are resolved.

Biodiversity and a precautionary approach

Biodiversity conservation and sustainable use of biological resources are characterised by the necessity of a precautionary approach: an approach related to the consideration of how human activities influence biodiversity, and to the question of how to avoid or minimise such impacts. The biodiversity concept, as mentioned earlier, not only relates to the protection of genes, species and habitats, but also to the protection of ecosystems, and thereby their processes, functions and services. Precaution is perhaps the driving force behind the concept, and as such it is reflected in the preamble of the Convention as conscious of the different values *and conscious also of the importance of biological diversity for evolution and for maintaining life- sustaining systems in the biosphere.* Also, the notion of the interests of future generations is in line with such an approach. As we cannot at present identify the needs and aspirations of future generations, a precautionary approach to biodiversity preservation (and sustainable use) seems to be the only way in which present generations can safeguard the interests of future generations.

A precautionary approach recognises the often significant degree of uncertainty in relation to the impact of various human activities on biodiversity. *The precautionary principle* is reflected in the preamble of the Biodiversity Convention, by

> "… noting also that where there is a threat of significant reduction or loss of biological diversity, lack of full scientific certainty should not be used as a reason for postponing measures to avoid or minimize such a threat, …"

In this sense, the principle is directed towards decisions on how to deal with a specific threat against biodiversity, i.e., a narrower expression of the precautionary approach. Specific procedures, e.g., risk assessment, are designed to clarify the degree of uncertainty, hence establishing the possibility of making an informed decision, balancing other involved interests against the uncertainty, or at least establishing a clear indication of how the uncertainty has been dealt with in reaching a decision. Other regulatory instruments may be used in trying to operationalise or implement the precautionary principle. Reversing the burden of proof has often been put forward as a means of doing this – that is, the "developer" has to prove or present sufficient evidence that no unacceptable risks (or uncertainties) are involved in the case in question.

A precautionary approach expresses a rationality based on virtues – that is, a (moral) responsibility to deliberate on the inevitable uncertainties of a case – but not an obligation to let uncertainty overrule all other considerations. Essentially, a precautionary approach requires a careful, open balancing of the different interests involved, at the same time emphasising the aspect of uncertainty. Openness and transparency are prerequisites for responsible decision-making. Risk assessments and other forms of assessment are designed not only to deal with uncertainty, but also to establish the basis for an open, transparent decision-making process. However, it may be argued that such instruments express a relatively methodological rationality – with the objective of clarifying impacts, risks and uncertainties by following certain steps or criteria for success laid down in advance. Nevertheless, they also express the virtues of deliberation, openness and willingness to consider new, alternative ways of solving the problems, i.e., one type of rationality does not exclude another on the long road to formulation and implementation of environmental policy objectives.

Biodiversity measures

In setting standards and targets, it is necessary to take into account the means available to achieve them (Sagoff, 1988, p. 220). This should imply, not only considerations of existing means – perhaps adopted in different pieces of legislation, but also reflections on the appropriateness of such means and on new means to be developed and possibly integrated into existing legal systems. The central elements must be considerations of how implementation can be carried out and of anticipated problems. Consequently, it is important to focus on counterproductive elements in existing legal systems and on unintended consequences of law and legal action (Dalberg-Larsen, 1989). One example could be considerations of how biodiversity measures fit into the traditions of private property rights.

Although it is not possible to make sharp distinctions between measures designed to secure biodiversity conservation, sustainable use and equitable sharing of benefits, since these three objectives are all interrelated; it is, however, possible to point out some characteristics relevant to the discussion of rationality. Here, different types of legal and regulatory initiatives are required, as well as different types of interest balancing – and, as a consequence, perhaps also different types of rationality.

Biodiversity conservation – in the narrow sense of species- and site-specific measures – to a large extent falls under traditional environmental law focusing on nature protection issues. The interests involved are species and habitat protection, recreational and aesthetic values, land-use interests and private property rights. The balancing of different interests can be more or less well defined. One example could be the U.S. Endangered Species Act (ESA), which to a certain degree narrows down the possibility of balancing different interests, as it offers "strict" protection (listing) of endangered species, although modified by cost-benefit analyses (Flournoy, 1993). It has been argued that the strict nature of the ESA may be counterproductive, as it disregards the responsibility of, and acceptance by, private property holders (Olson, 1996, p. 68). Furthermore, it has been argued that the species-by-species approach is inadequate in the pursuit of broader biodiversity objectives (Flournoy, 1993). Traditional species- and site-specific measures, such as listing, area designations, etc., tend to create conflicts with private property holders and other economic interests. Although compensation may be offered, it will seldom seem sufficient to the landowner. Therefore, alternative measures are being developed in order to appeal to the moral responsibility or the rational economic considerations of private property holders. Such initiatives include conservation or management agreements, incentives to promote conservation efforts, and the removal of disincentives (Preston, 1995). In art. 11, the Biodiversity Convention underlines the importance of adopting economically and socially sound measures as incentives for conservation and sustainable use of biological resources.

Sustainable use of biological resources implies another type of interest balancing, focusing on the integration of biodiversity considerations in sectoral decision-making, e.g., in the sectors of agriculture, fishery, tourism, transport etc. Adverse impact is inherently connected to the use of biological resources, and sustainable use is first and foremost a question of mitigating adverse impact in both the short and long terms. A precautionary approach is a key element in efforts to establish patterns of sustainable use of biological resources, including an ability to deal with the prevailing element of uncertainty. Open and informed decision-making is a central element in achieving sustainable use of biological resources. Thus the concept of sustainable use emphasises a rationality based on communication and reflection.

A number of different measures are designed to establish a procedural framework for such decisions. This includes environmental impact assessments, both at project and more strategic levels, risk assessments specifically designed to deal with uncertainty, public participation, access to justice, etc. Some of these measures or principles have been stressed in the Rio Declaration, and some of them also play a role in relation to the Biodiversity Convention. However, the requirements may have been rather vaguely formulated in the Convention, e.g., art. 14 on *appropriate* procedures for environmental impact assessment and public participation, where *appropriate*. According to art. 10(e), the contracting parties shall encourage co-operation between governmental authorities and the private sector in developing methods for sustainable use of biological resources. States are also obliged to promote and encourage understanding of the importance of, and the measures required for, biodiversity conservation (art. 13). The Convention further focuses on access to and use of genetic resources, and on biotechnology. These requirements can be seen as means of achieving the sustainable use of biological resources, but they are primarily related to the objective of the equitable sharing of benefits.

International conventions and modern declarations mirror the relationship between environmental protection and economic growth. The principle of *equitable sharing of benefits* arising out of the utilisation of genetic resources is central in the efforts to secure the balance between environment and economic development (Johnston, 1997, p. 260). Developing countries seek monetary assistance, and at the same time strain to secure effective institutions that can verify and ensure that the contracting parties comply with their obligations according to environmental contracts (Sands, 1994). These two characteristics have resulted in new trends in environmental law, where provisions laid down in the conventions provide for 'compensatory' funds to be made available to developing countries to enable them to meet certain 'incremental costs' of implementing their obligations, as well as providing for subsidiary authorities to verify compliance and implementation, *inter alia* the Global Environmental Facility (Johnston, 1996; Werksman, 1997).

Requirements for the equitable sharing of benefits play a central role in the Biodiversity Convention – reflecting the bargaining position of developing – and often biodiversity-rich – countries. Apart from financial mechanisms, equitable sharing of benefits refers to sustainable use of biological resources, technology transfer and control of the access to genetic resources. The latter may include use of intellectual property rights and contractual relationships (Johnston, 1997). A so-called Biosafety-protocol has been adopted under the auspices of the Convention (art. 19(3)). A core element in this protocol is *advance informed agreement* (AIA) – a parallel to the principle *of prior informed consent* (PIC).

The equitable sharing of benefits to some extent calls for economic responsibility in developing countries. Biodiversity protection becomes an economic interest, perceiving biological resources like other natural resources available for future exploitation. The balancing of interests in this respect may be said to reflect more traditional (economic) rational decision-making, in which the parties seek to reach a reasonable distribution of cost and benefits. The international requirements of *inter alia* the Biodiversity Convention seek to provide a framework for such agreements, with the purpose of creating incentives for biodiversity protection by securing the rights and interests of developing countries in particular.

Biodiversity and property rights – an example of the balancing of interest

Biodiversity regulation calls for a re-assessment of the traditional framework and principles of international as well as national law. One element of this might be the rethinking of the property rights concept – perhaps emphasising the aspects of responsibility, and focusing on an interest-based approach rather than a rights-based approach. In this respect, it is important that the management of natural resources is based on the recognition that degradation (incl. pollution) of the environment does not respect national boundaries or property-lines.

Ownership or *private property rights* can be regarded as one of the institutions that changes with the growth of civilisation. The concept of the legal right to land is based on the idea that land primarily has value, not as land, but as property, an idea that can be traced to John Locke, in his article "Second Treatise of Government" (Sagoff, 1988, p. 173). The concept of ownership is the common denominator for the following elements: the owner's possession or his primary right, which means a permanently protected freedom to make use of the object; the owner's competence, which is manifested in the power to transfer or lease the object, etc., and the demand for the owner's protection of the environment. The protection of property rights against interference from the authorities is primarily dealt with in constitutional law, in the principles of public law (e.g., the legality principle), in neighbour law and in tort law.

The constitutions of many countries stipulate that no one can be deprived of his private property rights, unless in accordance with the statutes, and unless full compensation is paid. Such rules are based on a liberal approach, as described above in 1.2. Expropriation law may require that any public law which lays down intensive or unusual restrictions on the right of private owners to dispose of their property must also compensate them for so doing. Constitutional protection, however, seldom prescribes absolute rights, and so-called general compensation-free regulation based on public interests (e.g., environmental protection) is well known in many countries.

Property rights play a central role in relation to the protection of biodiversity. Private property rights – i.e., the rights of landowners – are often referred to as one of the greatest obstacles in relation to biodiversity conservation. Meanwhile, another type of property right – namely intellectual property rights, referring to the rights of holders or creators of (indigenous) knowledge of biological resources – has been put forward in the Biodiversity Convention as a means of promoting biodiversity conservation. Furthermore, the Convention explicitly recognises the sovereign rights of States over their natural resources (art. 15). The latter two perceptions of property rights emphasise the responsibilities of property right holders, while the former focuses on the irresponsibility of private property holders. This paradox illustrates the fact that property rights include both elements, both of responsibility and possibly of irresponsibility – or short-term individual interests.

Traditionally, focus is placed on the *protection* of private property rights (and short term individual interests) rather than the *responsibilities* or moral obligations of landowners. It can, however, be questioned whether this perception is accordance with the rationale of the property rights concept and with legal definitions of private property rights. One example stressing both aspects is the German Constitutional notion of "eigentum verplichtet" (Michanek, 1997). Furthermore, the freedom of private property holders has traditionally been limited, at least to the extent that use of private property should not cause harm or nuisance to other persons, e.g., neighbours or other individual economic interests. Such limitations traditionally relate to pollution issues, but they may also concern nature protection issues, e.g., aesthetic or recreational interests, while they do not normally refer to non-economic nature protection interests. Environmental protection law has supplemented neighbour law with limitations regarding the protection of public health against pollution.

Where the line should be drawn between inviolable private property rights and compensation-free private property interests is a matter of debate in many countries. In a Scandinavian context, it appears that costs related to restrictions in future land use must be borne by the landowner, while restrictions in existing land use are more likely to lead to compensation (Michanek, 1997). But the borderlines are somewhat unclear in this respect. In the US, a central distinction is made between the exercise of police power and the exercise of eminent domain (taking land for public uses). The former is exercised for the purpose of controlling public nuisances and without compensation (Carlton, 1986). Although the legal and cultural context may provide various criteria, restrictions directed towards biodiversity protection are difficult to categorise.

The challenge of environmental law – and biodiversity law in particular – seems to be the focus on the inherent element of responsibility. A conflicting approach – setting environment and private property fundamentally at odds – cannot endure, even if laws protecting biodiversity could be strictly enforced (Olson, 1996, p. 69). Although there is a core of private property rights which is respected in most legal systems, private property is also an interest which must be properly deliberated on in an interest-balancing process. Accordingly, it is necessary to organise the legal system in a way that enables a proper balance to be struck in each case.

CONCLUDING REMARKS

The increasing complexity of environmental law – and "biodiversity law" in particular – suggests that innovative efforts and the development of new measures to pursue established goals are essential. The kind of rationality central to such a process can, of course, be the subject of discussion. The answer is probably that a mix of different types of rationality is needed. As has been pointed out, however, inherent elements of the overall objectives of sustainable development and biodiversity protection assert the need to provide the necessary basis for sound, balanced decision-making. In doing so, a number of different interests need to be taken into account. Traditional economic rationality and economic valuation techniques may not be sufficient basis for such decision-making, as a number of more or less uncertain presumptions are used. Neither will a narrow norm-orientated or goal-orientated rationality provide a sufficient basis, as both norms and goals are rarely precise or unambiguous. In order to provide a basis for representation and influence by all interests involved, there appears to be a need to supplement these kinds of rationality with one based on the concepts or virtues of deliberation and communication. In this respect, the legal system may be part of the institutionalisation and formalisation of a rationality of norms based on such ideas. This could be done by strengthening the legal status of environmental principles, or by introducing new regulatory means or measures.

Sustainable development and biodiversity protection require a case-by-case balancing of the different interests and values involved. In many cases, such rationality may be reflected in administrative practice in relation to environmental law. At least in Denmark, the notion of consensus (and a negotiating process) has emerged, both in relation to legislative work and to administrative practice. This, however, is not always the case, and the focus will not always be on the importance of different legal traditions and factors in influencing implementation of the objectives. A consensus approach may serve as a cover for scientific uncertainty, and openness is not necessarily part of the process. Parties involved in negotiations may be restricted to a narrow group of representatives from authorities and powerful organisations (both environmental and business). Trade-offs agreed on in such a narrow group do not necessarily reflect the interests of citizens, of future generations, or of nature.

Thus, legal principles of public participation, access to justice, access to information and legal certainty -i.e., procedural guarantees, are central in relation to the involvement of all (relevant) interests. A real consensus or communicative approach aims at securing a broad understanding and acceptance of not only the objectives – but also of the instruments necessary to implement them. It must be acknowledged that environmental objectives can seldom be achieved solely by legislative or regulatory initiatives. Moral considerations and a precautionary approach must be part of "legal thinking" and of everyday life. Clearly, international, regional, national and local initiatives are still needed – but at the end of the day it is the rationality of the individual as a citizen in an environmentally sensitive community that really matters. Establishing a legal framework based on the above-mentioned ideas plays a central part in promoting a similar rationality in decision-making by authorities and the individual citizen. The challenge to the legal system is therefore to express means and ends in a realistic and balanced way. As expressed by Robinson (1992) ".. *laws exist in order to protect, encourage and to stimulate the law abiding and to guide or influence citizens as to the course of action which they should adopt"*.

REFERENCES

Anker, Helle T. & Ellen Margrethe Basse (ed.) (2000). *Land Use and Nature Protection. Emerging Legal Aspects.* Copenhagen: Djøf Publiching.

Basse, Ellen Margrethe (1987). *Miljøankenævnet* (The Environmental Appeal Board), Copenhagen: Gad.

Basse, Ellen Margrethe (1996). Public Environmental Law in Denmark. In: Seerden, R. & Heldeweg, M.(eds.), *Comparative Environmental Law in Europe. An Introduction to Public Environmental Law in the EU Member States*, Antwerpen: MAKLU.

Birnie, Patricia W. and Boyle, Alan E. (1992). *International Law & the Environment.* Oxford: Clarendon Press.

Bowman, Michael (1996). The Nature Development and Philosophical Foundations of the Biodiversity Concept in International Law. In: Bowman, M. & Redgwell, C. (eds.), *International Law and the Conservation of Biological Diversity*, London: Kluwer Law International.

Callicott, J. Baird (1986). On the Intrinsic Value of Nonhuman Species. In: Norton B. (ed.), *The Preservation of Species, Princeton.* New Jersey: Princeton University Press.

Carlton, Robert L. (1986). Property Rights and Incentives in the Preservation of Species. In: Norton, Bryan G.(ed), *The Preservation of Species*. Princeton, New Jersey: Princeton University Press, 1986.

Dalberg-Larsen, Jørgen (1989). On the Evolution of Law and Unintended Consequences. In: Arnio, A. & Tuori, K. (eds.), *Law, Morality and Discursive Rationality*. Helsinki.

Dalberg-Larsen, Jørgen (1977). *Retsvidenskaben som samfundsvidenskab*, part II. (Legal Science as Part of Social Science).

Dalberg-Larsen, Jørgen (1977). Four Phases in the Development of Modern Legal Science. *Scandinavian Studies in Law*, Vol. 23.

Flournoy, Alyson (1993). Beyond the "Spotted Owl Problem": Learning from the Old-Growth Controversy. *Harvard Environmental Law Review*, Vol. 17, No.2, pp. 261-332.

Fuller, Lon L. (1964). *The Morality of Law*. New Haven: Yale University Press.

Graver, Hans Petter (1989). Rationality and the Development of Law. In: Aulis Arnio & Karlo Tuori (eds), *Law, Morality, and Discursive Rationality*, Helsinki.

Habermas, Jurgen (1992). *Moral Consciousness and Communicative Action*, translated by C. Lenhardt and S. Weber Nicholsen. Oxford: Polity Press (paperback), Blackwell Publishers.

Hydén, Håkan (1984). *Ram eller lag. Om ramlagstiftning och samhällsorganisation*. (Framework or Statute. On Framework Legislation and the Organization of Society). Stockholm.

Hydén, Håkan (1986). Sociology of Law in Scandinavia. *Journal of Law and Society*, Vol. 13.

Johnston, Sam (1997). North-South Tensions Within the Convention on Biological Diversity: A Case Study. In: Basse, E.M. (ed.), *Environmental Law. From International to National Law*. Copenhagen: GadJura.

Johnston, Sam (1996). Financial Aid, Biodiversity and International Law. In: Bowman, M. & Redgwee, C. (eds.), *International Law and the Conservation of Biological Diversity*. London: Kluwer Law International.

Koester, Veit (1997). The Biodiversity Convention Negotiation Process. In: Basse, E.M. (ed.), *Environmental Law. From International to National Law*. Copenhagen: GadJura, 1997.

MacCormick (1978). *Legal Reasoning and Legal Theory*. Oxford: Clarendon Press, reprinted in 1995.

Mann, Howard (1995). Comments on the Paper of Philippe Sands. In:

Lang, W. (ed), *Sustainable Development and International Law*. London: Graham & Trotman.

Merrymann, John Henry, Clarke, David S. and Haley, John O. (1994). *The Civil Law Tradition: Europe, Latin America, and East Asia*. Charlotteville, Virginia: The Michie Company Law Publisher.

Michanek, Gabriel (1997). National Protection of Biological Diversity. In: Basse, E.M.(ed), *Environmental Law. From International to National Law*. Copenhagen: GadJura.

Moe, Mogens (1995). Environmental Administration in Denmark. *Environmental News No. 19*. Copenhagen: Danish Environmental Protection Agency, Ministry of Environment and Energy.

Norton, Bryan, G. (1998). Ecology and freedom: sustainable opportunities in multi-generational communities. In: Arler, F & Svennevig, I. (eds.), *Cross-Cultural Protection of Nature and Environment*. Odense University Press.

Norton, Bryan G. (1991). *Toward Unity Among Environmentalists*. New York: Oxford University Press.

Olson, Todd G. (1996). Biodiversity and Private Property: Conflict or Opportunity? In: Snape, W.J. (ed.), *Biodiversity and the Law*.

Perrings, C.A. et al. (1995). *Biodiversity Conservation. Problems and Policies*. Dordrecht: Klüwer Academic Publishers.

Preston, Brian (1995). The Role of Law in the Protection of Biological Diversity in the Asia-Pacific Region. *Environmental and Planning Law Journal*, Vol. 12, No. 4, pp. 264-277.

Robinson, Nicholas (1992). Caring for the Earth – A Legal Blueprint for Sustainable Development. *Environmental Policy and Law* 22, p 22-23.

Ross, Alf (1953). Om Ret og Retfærdighed. Copenhagen. (translation: On Law and Justice, London, 1958).

Sacco, Rodolfo (1991). Legal Formants: A Dynamic Approach to Comparative Law. *The American Journal of Comparative Law*, Vol. 39, pp. 343-401.

Sagoff, Mark (1988). *The Economy of the Earth. Philosophy, Law and the Environment*. Cambridge: Cambridge University Press.

Sands, Philippe (1994). *Greening International Law*. New York: New Press.

Sands, Philippe (1995). International Law in the Field of Sustainable Development: Emerging Legal Principles. In: Lang, W. (ed.), *Sustainable Development and International Law*. London: Graham & Trotman.

Skirbekk, Gunnar (ed.) (1994). *The Notion of Sustainability and its Normative Implications*. Scandinavian University Press.

Snape, William J.(ed.) (1996:). *Biodiversity and the law*. Island Press.

Stone, Christopher D. (1972). Should Trees Have Standing – Toward Legal Rights for Natural Objects. *Southern California Law Review*, Vol. 45, pp. 450-501.

Teubner, Günther (1983). Substantive and Reflexive Elements in Modern Law. *Law and Society Review*, Vol. 17, No. 2, pp. 239-285.

Werksman, Jacob (1997). The Global Environment Facility. In: E.M. Basse (ed.), *Environmental Law. From International to National Law*, Copenhagen: GadJura.

CHAPTER 7

When Choice of Means Undermines the Goals: Rationality from a Psychological Perspective

John Thøgersen and Tommy Gärling

INTRODUCTION

EVERY TIME WE hear bad news about the progressive deterioration of the environment, we ask ourselves whether or not it is inevitable. The late Swedish Prime Minister, Olof Palme, often expressed optimism, based on his firm belief that people are rational. He implied (1) that the environmental problems we face today are solvable, (2) that knowledge of how to solve them is available or will be made available, and (3) that when such knowledge exists, it will be used to solve the problems. Taking a psychological perspective on rationality in this chapter, we see less reason to be optimistic. Even though many environmental problems are solvable and knowledge about solutions is available, there appear to be major obstacles to implementing the solutions.

Knowledge about the future is always uncertain. Therefore, different opinions exist about what is a rational course of action. Even worse, there is not always consensus about the meaning of rational. We will briefly discuss this issue from a psychological perspective.

The concept of rationality introduced in rational choice theory (Tversky & Kahneman, 1986) may serve as a standard. Relative to this standard, it is possible to assess the rationality of both societal and consumer decisions having environmental consequences. Extensive research on human decision-making (Camerer, 1995) has, in fact, done exactly this. Thus, we have a firm basis for stating that decision-making frequently departs from rationality. Any analysis of the means of solutions of environmental problems needs to take this into account. Relying on the belief that rational choice theory is invariably an adequate description of human decision-making, either at a societal or at an individual level, ignores what is currently known. We will briefly review this knowledge.

As others have argued and we too, will argue, rational choice theory describes a logic of choice, rather than a psychology of value. It thereby diverts attention from the important issue of how and why different values are prioritised. Psychological research addressing this question has revealed what these values are in the minds of ordinary people, how they derive from the universal human requirement of adapting to reality, and how the various values are prioritised in different cultures. A theory of rational choice must include a theory of values.

An underlying assumption of at least some versions of rational choice theory is that self-interest alone motivates human action. Consequently, it is believed that environmental problems in part arise because self-interest is not restrained (Hardin, 1968). However, that self-interest motivates all human action has been challenged on empirical grounds (Bateson, Dyck, Brandt, Bateson, Powell, McMaster, & Griffitt, 1988; Caporael, Dawes, Orbell, & Kragt, 1989). In particular, the self-interest assumption ignores the role group living may have played in the evolution of the human species (Gardner & Stern, 1996). Again from a psychological perspective, we will briefly discuss the conflict between individual and collective rationality, followed by a review of research demonstrating that there are a number of factors that may promote collective rationality in human decision-making.

Using the discussion of the issues outlined above as a backdrop, we will illustrate how existing consumption-related environmental problems can be attributed to deficiencies and conflicts of human rationality in decision-making. After that, we discuss to what degree various means proposed for solving the problems are consistent or inconsistent with what we know about human decision-making.

LIMITS TO RATIONALITY

Defining rationality

In psychological research on decision-making, rationality is defined with reference to rational choice theory (Tversky & Kahneman, 1986). In this theory, rationality refers to the internal consistency of preferences, without regard to what these preferences actually are (Levi, Cook, O'Brien,

& Faye, 1990). The bulk of empirical research has examined the circumstances under which principles such as maximization of expected utility, presupposing consistent preferences, are followed (Dawes, 1998). To account for the discrepancies observed, several alternative principles have been proposed (e.g., Kahnman & Tversky, 1979; Hogarth & Einhorn, 1990). In an analysis of cognitive requirements for rational decisions, Frisch and Clemen (1994) note that rational choice theory implies (1) that a choice of action should be based entirely on the anticipated consequences of that action, (2) that the consequences of the action should be accurately anticipated or structured, and (3) that trade-offs should be made between different consequences of an action. In this way the axioms of rational choice theory are translated into assumptions about how decision-makers process information, making it possible to investigate whether decision-making is consistent with these assumptions. Based on such investigations, and closely related to the notion of bounded rationality (Simon, 1982), alternative theories have been proposed on the basis of a psychological cost-benefit principle (Payne, Bettman, & Johnson, 1990).

If internal consistency of beliefs and preferences is the sole criterion of rationality, then the situation may be easily rectified. As Kahneman (1994, p. 19) notes, "... if subtle inconsistencies are the worst indictment of human rationality, there is little to worry about." However, when the set of conditions for rationality provided by the standard theory of choice "allow many foolish decisions to be called rational" (Kahneman, 1994, p. 32), rationality becomes a dubious yardstick for the quality of decision-making. Attempts to reconcile observed preferences with rationality by adopting a more permissive definition of rational choice further weaken the rationality concept in this respect. Kahneman argues that decision theory should move in the opposite direction, making the criteria of rationality more restrictive, among other things by adding substantive criteria of rationality to the logical standard of coherence (Sen, 1990). For instance, he argues that maximizing *expected* utility (one among many goals individuals may entertain) should not be sufficient for a choice to be termed rational. Only choices that also maximize *experienced* utility deserve this designation, implying that rational decision-makers are able to correctly foresee the outcomes of their choices, as well as their preference for future outcomes, at the time of making them. Experimental research suggests that people lack skills in both these tasks (Kahneman, 1994).

An illuminating related discussion of this issue is found in Brennan

(1990). Brennan notes that a nonstandard version of rational choice theory exists, which makes assumptions about rationality of ends, not only of means. Self-interest is a general motivational principle which is evoked in this context. However, like many others, Brennan himself seems to believe that a choice theory encompassing motivational components is not feasible. We do not fully agree. In fact, if a universal motivational theory delineates what people value most, such a theory would certainly be a most valuable supplement to the logical decision rules of rational choice (Dietz & Stern, 1995; Edney, 1980; Etzioni, 1988). The questions this raises are whether such a theory exists and, if it exists, how it may be used to assess the rationality of behaviour.

A motivational theory which has been extensively empirically tested in cross-cultural research is the psychological value theory proposed by Schwartz (1992, 1994; Schwartz & Bilsky, 1987, 1990). In this theory, values are defined as beliefs pertaining to desirable end states or modes of conduct that transcend specific situations, guide choices of actions, and can be ordered by importance to form a hierarchy of value priorities. A typology is derived from the assumption that values refer to a finite number of motivational concerns originating from the requirement of coping with reality: individual needs, social needs, and social institutional needs, respectively. Through socialization, individuals learn to think of these motivational concerns as conscious values, to use terms to communicate these values, and to ascribe different degrees of importance to them.

Even though the contents and structure of values may be universal (Schwartz & Sagiv, 1995), there are clearly both cultural and individual differences in value priorities. Furthermore, the value structure implies that trade-offs are inevitable. For instance, collectivistic values cannot be prioritised at the same time as individualistic values. Thus, further research is needed to find out if it is possible to disclose universal trade-offs. Obviously, standard rational choice theory only helps to prescribe how to make *consistent* tradeoffs.

Descriptive theories of rational decision making

Revisions of rational choice theory

It is obvious to most decision-makers that violations of a number of the assumptions made in expected utility theory are irrational. Most decision-makers nevertheless frequently violate these assumptions. Why do they do that? Take, for instance, violations of the cancellation assumption or 'sure thing' principle (Savage, 1954; Von Neumann & Morgenstern, 1944). It states that if one option is preferred over another, no matter what occurs (or has occurred), knowledge of what occurs should have no influence on the decision. Few would contest that this is rational. Yet, violations are frequently registered. For example, in a study by Tversky and Shafir (1992), students wanted to postpone a decision concerning a vacation trip until they knew if they failed or passed an exam, although they would go in any case.

An important objective of current research into why decision-makers violate obvious principles of rationality is to provide detailed information about the violations themselves (Tversky & Kahneman, 1986), in order to revise the rational theory so that it accommodates empirical findings. A good example of such a revision is prospect theory (Kahneman & Tversky, 1979; Tversky & Fox, 1995). This theory is consistent with several known violations of cancellation. It is likewise consistent with violations of the assumption of descriptive invariance, which implies that utilities of outcomes should not be influenced by how they are described. An example is that outcomes framed as losses (e.g., a 50% chance of dying from a medical treatment) are evaluated more negatively than the same outcomes framed as gains (a 50% chance of recovery from a medical treatment). Prospect theory assumes that outcomes are first framed as gains or losses in an editing phase (selecting a subjective zero or reference point of the utility function), then evaluated according to an S-shaped utility function (convex and steeper for losses, concave for gains), and finally multiplied by decision weights and summed over outcomes. The editing phase also entails integration or segregation of outcomes of previous decisions, for instance, whether or not to ignore a sunk cost in evaluating a current outcome (Gärling, Karlsson, Romanus, & Selart, 1997). Furthermore, as illustrated in Tversky and Kahneman (1986), editing with the aim of cognitively simplifying the options (e.g., combining outcomes with the same utility but different probabilities), may violate the assumption of dominance because dominance of one option over another is made nontransparent.[21] In sum, the important differences from expected utility theory are

21. It is assumed that transparent forms of violations of dominance are discovered in the editing phase of prospect theory, witnessing to the fact that this principle, implying no tradeoffs, is so obviously rational. However, in this example, discovering dominance is in conflict with the goal of cognitive simplification.

the assumption of an editing phase; a different shape of the utility function, more consistent with existing knowledge about perceptual processes; and the assumption that utilities are multiplied by decision weights rather than by probabilities. The theory is consistent with the fact that human beings have a limited cognitive capacity. The decision task is accordingly cognitively simplified in the editing phase. Thus, limits on cognitive capacity are one reason for departures from rationality. The shape of the utility function reflects the principles that the perceptual system reacts to changes rather than to steady states, is influenced by expectations, and places heavier weight on potential losses than gains.

A distinction is often made between riskless decisions, decisions under risk (when probabilities of uncertain outcomes are known), and decision under uncertainty (when probabilities of uncertain outcomes are not known or incompletely known). Dawes (1998) and others have questioned whether decisions are ever riskless. Many consumer choices would be classified as riskless. However, even though it is possible to know with certainty what product is purchased, the effects of using it may be uncertain. It may similarly be questioned whether outcome probabilities are ever known, except in a lottery (and then only by the owner). An important research focus has therefore been how judgments of probability are made. According to rational choice theory, such judgments should not violate the axioms of probability theory. However, they invariably do (Kahneman, Slovic, & Tversky, 1982). In prospect theory generalized to decisions under uncertainty (Tverky & Fox, 1995), it is assumed that judged probability is related to probability so that probabilities close to 0 are underestimated (actually "rounded off" to 0). Small probabilities farther from 0 are overestimated. Along the remaining sections of the continuum, probabilities are underestimated, except close to 1, where they are again overestimated ("rounded off" to 1). The decision weights by which utilities are multiplied show a similar relationship to judged probabilities. These assumptions, in conjunction with the assumption of an S-shaped utility function, account for an empirically verified four-fold pattern of risk seeking (which is not consistent with maximization of expected utility): People seek risk (choose a risky option over a sure option with the same expected utility) when there is a small probability of winning a large amount (e.g., buying lottery tickets) or a large probability of losing a small amount; however, they avoid risk when there is a large probability of winning a small amount (although, especially in this case, there are important individual differences which should perhaps not be ignored; see Lopes, 1987) or a small probability of losing a large amount (e.g., buying insurance).

Still another violation of the expected utility principle is that utility is not always independent of probability, that is, if outcome A with utility u_A is preferred over outcome B with utility u_B, it does not necessarily follow that pA is preferred over pB where p denotes the probability of the outcomes. Since the direction of the violation is that p increases with u (more desirable outcomes are judged to be more likely, less desirable outcomes less likely), the phenomenon is labelled an outcome-desirability or wishful-thinking bias (Budescu & Bruderman, 1995). Venture theory (Hogarth & Einhorn, 1990) differs from prospect theory in its additional assumption that the decision weights are not only related to probability, but also to the utility of the outcome. Furthermore, the effect of the latter increases with the uncertainty or ambiguity of the judged probability. A decision-maker is assumed to adjust the decision weight from an initial anchor (which may be a judged probability) by imagining possible negative and positive outcomes. The less constraining the information is, the more extensive adjustments are. There is also more room for wishful thinking.

Extensions of rational choice theory

Other reasons for not making rational decisions emerge from a third research focus: examination of the decision strategies that expert decision-makers as well as lay people use (Payne et al., 1990). A decision strategy is defined as a sequence of mental operations which transform an initial state of knowledge into a final state of knowledge where the decision maker considers the decision problem solved (i.e., he or she knows which option to choose). A large number of decision strategies have been found to be used. Decision strategies may be characterized with respect to how accurate and effortful they are. An example of an inaccurate decision strategy which requires little effort is retrieving a previous decision from memory. An accurate decision strategy requiring much effort is the expected-utility decision strategy (first multiply utilities by probabilities, then sum the products for each option, and, finally, chose the option with the highest sum).

There are two main insights from studies of decision strategies (Ford, Schmitt, Schechtman, Hults, & Doherty, 1989; Payne et al., 1990). First, verifying Simon's (Simon, 1982) principle of bounded rationality, lay people (but sometimes more frequently experts, as reviewed by Camerer & Johnson, 1991,) use satisfying decision strategies. Thus, people often choose options which they find satisfactory, without considering whether better options exist. Another characteristic of decision strategies people use is that they are noncompensatory, i.e., unlike the expected-utility decision strategy, they avoid value trade-offs. Although there is a cost in terms of lower accuracy, computer simulations of noncompensatory decision strategies (Payne, Bettman & Johnson, 1988) have indicated that the cost may be marginal under many circumstances. Under other circumstances, such as time pressure, noncompensatory decision strategies may even be more accurate than expected utility.

Second, the use of a decision strategy is the outcome of a decision on how to decide. Several factors, including characteristics of the decision maker (e.g., prior knowledge, cognitive ability), the decision task (e.g., the number of options, available information about the options), and the social context (e.g., the need to account for the decision, social pressure) have been shown to influence the choice of decision strategy (Ford et al., 1989). At a more detailed level of analysis, it has been proposed that the choice reflects a trade-off between costs and benefits (Payne et al., 1990). In different environments, the accuracy of a particular decision strategy varies, as does the effort required for its execution. The adaptive decision-maker is successful in choosing the decision strategy with the highest attainable accuracy for the least effort. Knowledge of how such metadecisions (on being rational or not) are made is currently incomplete. They may sometimes be characterized as top-down processes, i.e., based, for instance, on information about the decision task and its importance. However, there is also evidence for bottom-up processes: A decision-maker may choose, or even construct, a decision strategy while being engaged in the decision-making.

In addition to cognitive limitations and perceptual principles, people who make decisions are sometimes not rational because they do not think it is worth the effort. Thus, they employ a broader utility concept, additionally including the costs of information search, mental effort, time spent, and so forth.

INDIVIDUAL VERSUS COLLECTIVE RATIONALITY

Defining collective rationality

Collective rationality is the extent to which groups of individuals effectively pursue common (collective) goals or values (Etzioni, 1988). At the individual level, collective rationality implies that group members contribute to the common quest in a manner appropriate (necessary and sufficient) for achieving common goals or values. Hence, collective rationality implies that group members sometimes give higher priority to common than to individual goals or values. In (primarily voluntary) groups with shared goals and values, there will seldom be a conflict between individual and collective rationality. However, there almost certainly will be in groups formed by force of powerful others or of common destiny, e.g., common dependence on or desire for a certain resource, (e.g., Gardner & Stern, 1996; Ostrom, 1990).

Even in cases with shared values and goals, an important source of conflict between individual and collective rationality is that individual rationality prescribes that the individual not contribute more than necessary to achieve the common goal. For an individual who believes that others will make sufficient contributions to attain the common goal[22], it is individually rational not to contribute at all, that is, to be a "free-rider." However, if many share this belief, and they all follow the individually rational course of action, the common goal will not be attained.

The conflict between individual and collective rationality may be illustrated with an example from game theory, the *N*-person game[23] (Komorita, 1976), applied to the problem of managing a scarce renewable resource. In this game each of *N* players has a choice of how much to use of a freely accessible resource. If the total amount used exceeds an upper limit, the resource will be depleted. On the other hand, because the resource is renewable, it will be accessible in the future if the total amount does not exceed this limit. However, since the resource is scarce, it is not possible for everyone to use as much as they want. As long as the resource is freely accessible, is there any reason why the

22. Assuming for the moment that the goal can be achieved without everyone contributing, equally, or according to ability.
23. The *N*-person game is a generalization of the well-known prisoner's dilemma game (Luce & Raiffa, 1957) for more than two players.

players should restrain themselves? Would information about the collective consequence (that the resource will be depleted in the future) make them do so? It might instead motivate hoarding. What about knowledge about the other players? Is there any such knowledge which would restrain the players; for instance that the other players are relatives, close friends, associates, or simply people similar to oneself? If such knowledge should lead to the conviction that the other players are trustworthy and likely to restrain themselves, a player may even use more of the resource than he or she would otherwise do, because more is available for him or her. Will players restrain themselves if depletion of the resource occurs in the future, when they are no longer around, so that the benefits of restraining from using the resource will accrue to others? Does it matter who these others are?

A strong "Hobbesian" tradition in the social and behavioural sciences assumes that, due to egoism, people will certainly not show restraint in social dilemmas if they are not coerced to do so (Mansbridge, 1990). This assumption is, however, not confirmed by experimental research. As soon as people have been familiarized with the rules of the game, a fairly large proportion of them restrain themselves, even though they seem to gain nothing from doing so (Caporael et al., 1989). Such experimental findings are reviewed in slightly more detail in the following.

Factors promoting collectively rational decision making

Structural factors

In the *N*-person game, the outcome of each player's choice depends on the choices made by the other players. Individual players are better off if they choose to use as much as they want of the resource, if a sufficient number of others do not. Obtaining this outcome is therefore an incentive to use the resource. However, fear of being a "sucker" (Wiener & Doescher, 1991) may be an equally strong or stronger incentive to use the resource. Players become suckers if they choose to restrain themselves, but a sufficient number of others do not, and the resource is depleted. Not surprisingly, the size of the potential gain from unrestrained use (the outcome structure of the *N*-person game) has strong consistent effects on restraint (Van Lange et al., 1992). However, whether greed or fear of becoming a sucker is the more important reason seems to vary (Bruins, Liebrand, & Wilke, 1989). Furthermore, whether more restraint can be obtained by rewarding restraint or punishing nonrestraint is an unsettled issue (Komorita & Barth, 1985).

Limiting free access to a resource is another means of producing restraint. The important issue here is when such a change is accepted. The answer depends on whether the resource is being overused, whether access is unfair (that some, with no justification, have more access than others), and the way in which free access is limited. In general, privatisation seems to be preferred to an institutional solution (Samuelson & Messick, 1995), possibly because it infringes less on freedom. Other factors which have been found to influence acceptance of means of limiting free access are efficiency, fairness, and self-interest (Samuelson, 1993).

Situational factors

Without being as coercive as limits on free access, the social situation has been found to have profound effects on how much individuals restrain themselves in the *N*-person game, even when each player has free access to the resource. Identifiability, group identity, feelings of personal responsibility, and perceived efficacy are factors associated with the social situation that have been shown to increase restraint (Van Lange et el., 1992). Other studies (Orbell, Kragt, & Dawes, 1988) have demonstrated the important role of communication between group members.

Interaction among group members may also increase the salience of norms or the consequences of norm violations. A norm is defined as an expectation held by an individual about how he or she ought to act in a particular social situation (Schwartz, 1977). It is a social norm if it must be enforced by the threat of punishment or promise of reward; otherwise, it is an internalised, personal norm or guiding principle for how to act.[24] In the context of *N*-person dilemmas, it would be of interest to identify behaviour-prescribing norms that directly or indirectly enhance collective rationality. Kerr (1995) exemplifies with the norms of commitment, fairness, and reciprocity. As noted above, mutual commitments, or social contracts, may be the most important reason why group discussions enhance restraint (Kerr & Kaufman-Gilliland, 1994). An equal share is usually accepted as a fair rule for how much each may use of a resource (Wilke, 1991). Justifable asymmetries, such as some being more needy than others, will make equity the fairness norm (Budescu,

---

24. The close relationship to instrumental values or desirable modes of conduct (Schwartz, 1992) should be noted.

Rapoport, & Suleiman, 1990; Van Dijk & Wilke, 1995). In iterated *N*-person games, consistent restraint seems to invite exploitation (Pruitt & Kimmle, 1977). However, reciprocating both restraints and nonrestraints, as in the tit-for-tat strategy (Axelrod, 1984), is very powerful in eliciting restraints. Other research suggests a correlation between own restraint and how much others are believed to restrain themselves, although not necessarily a causal relationship (e.g., Messick et al., 1983).

A related explanatory concept is role, defined as a set of norms; for instance, roles may trigger social responsibility. A particularly important finding is that acting as an elected leader increases restraint in *N*-person games (Messick et al., 1983), although the leader may feel justified in taking more for him- or herself.

Individual factors

As concluded by Kerr (1995), there may be little point in focusing on situation-specific norms. However, general internalised norms, such as those exemplified above, are likely to have explanatory power. Such norms may explain why individuals voluntarily restrain themselves, and may also contribute to an understanding of individual and cultural differences. Eventually it may be possible to disentangle the complex influences of socialization processes, which make people more likely to restrain themselves in the numerous "*N*-person games" they encounter throughout their lives.

In a related vein, research has investigated individual differences in social value orientations. An impetus for this research is the recurrent finding that between 40% and 60% of individuals refrain from using a resource, even though they are anonymous (e.g., Caporael et al., 1989). Do these individuals differ from those who do not restrain themselves under the same circumstances? A test, labelled the decomposed game (Kulhman & Marshello, 1975; Liebrand & McClintoch, 1988), has been used to classify individuals as "prosocials", "proselfs", or "competitors". In contrast to proselfs and competitors, in *N*-person games, prosocials place more value on others' outcomes, restrain themselves more, and expect others to do the same (Van Lange, Liebrand, Messick, & Wilke, 1992).

In a recent study, Gärling (1999) found a difference between prosocials and proselfs with respect to the priority they assign to collective values such as equality and social justice, whereas there was no difference in priorities assigned to benevolence values such as true friendship and good social relations. This finding highlights the relationship that may exist between social value orientations and internalised norms.

An N-person game is characterized by strategic or social uncertainty about how others will act, as well as, in many cases, uncertainty as to the size of a resource (Biel & Gärling, 1995). An important additional factor affecting restraint is therefore expectations as to whether or not others will restrain themselves (Pruitt & Kimmle, 1977). Thus, even though individuals goals are to restrain themselves, they may not do so unless they believe a sufficient number of others will. As noted, prosocials whose goal is to restrain themselves may believe that others will also do so. People also seem to vary in their readiness to trust others; "low trusters" experience a higher degree of social uncertainty than "high trusters" (Yamagishi, 1988).

At the individual level, lack of knowledge of the exact size of the resource may also offset restraint (Messick & McClelland, 1983). This has more recently been demonstrated in several experiments showing that uncertainty with regard to the size of the resource leads to outcome-desirability bias (Gärling, Gustafsson & Biel, 1998; Rapoport, Budescu, Suleiman, & Weg, 1992).

In summary, in addition to unconstrained self-interest, uncertainty or ignorance about others or the resource appear to constitute major reasons for people failing to show restraint in experimental N-person dilemmas. In real-life resource or commons dilemmas, people are often neither well informed, nor do they necessarily trust others. Some people may therefore deliberately overuse, pollute, or fail to contribute to valuable resources although they prioritise collective goals and values.

PRINCIPAL MEANS OF PROMOTING RESTRAINT

A classification of means

Means of promoting restraint in social dilemmas can be classified in a variety of ways. A widely used classification is the dichotomy of *structural* (altering the objective features of the decision situation) and *individual* (influencing relevant attitudes and beliefs) approaches (Messick & Brewer, 1983). Structural means are generally understood to be more coercive than individual means. However, this is a simplification. Not all structural solutions to real-life social dilemmas involve coercion, whereas some individual-psychological solutions may. Zaltman (1974) suggests classifying behaviour-change strategies into four categories, characterized by a decreasing level of coercion: (1) power, (2) persuasion, (3) re-education, and (4) facilitation. Both the most (power) and the least (facilitation) coercive strategy in Zaltman's classification involve altering the objective features of the decision situation. "Persuasion" covers the use of incentives and biased information (i.e., both structural and individual-psychological means), whereas re-education involves presenting all relevant information and stimulating reflection in order to enable the individual to take a more informed stance on issues.

Both of these classifications may be criticized for omitting or unduly de-emphasizing behaviour-change strategies that have historically proven their effectiveness. Ophuls (1973) proposed a classification of means into four basic types, which he claims cover all historic attempts to promote prosocial behaviour: (1) government laws, regulations, and incentives; (2) education; (3) informal social processes operating in small social groups and communities; and (4) moral, religious, and/or ethical appeals. The social and normative means included in Ophuls' classes 3 and 4 are easily overlooked when using the former two classifications. However, they have played an important role in solving many real-life social dilemmas (Gardner & Stern, 1996; Ostrom, 1990).

Intended effects and side effects

Environmental regulation aims to promote environmentally friendly behaviour by making the environmentally friendly option seem more individually rewarding, by convincing the individual that choosing this option is the right thing to do in spite of conflicts with his or her own short-sighted interests, and/or by facilitating environmentally friendly behaviour (by providing instructions, equipment, or other prerequisites). Threats of fines or jail, and taxes and subsidies are common means of making environmentally friendly behaviour more individually profitable compared to less environmentally friendly alternatives. Preferential treatment of individuals who behave in an environmentally friendly way (e.g., carpool lanes) serves the same purpose. Advertising and support for local community organizations are examples of means that have been used to inform individuals of the long-term environmental consequences of everyday activities and of how they can help avoid negative consequences. In practice, means are often used in combination, as when new laws or subsidies are advertised, or when instructions and feedback are combined with new opportunities for alternative action.

All regulatory means have their strengths and weaknesses. Besides their intended effect, they may have positive and negative side effects, usually unintended and unanticipated. To complicate matters even further, both intended effects and unintended side effects probably depend as much on the combination of means as on the individual means themselves (Gardner & Stern, 1996).

Policy-makers generally seem to trust coercive means most, probably because of their alleged ability to force restraint (Hardin, 1968; Olson, 1965). In addition to their forcefulness, coercive means may have a number of positive side effects that help fulfil their mission. However, negative side effects may sometimes more than outweigh positive, intended and unintended, effects. Non-coercive means generally have fewer negative side effects, but practical experience strongly indicates that they will not do the job alone if there are strong incentives to, or barriers against, showing restraint (Gardner & Stern, 1996; Klandermans, 1992).

Coercive means

Positive side effects. Besides providing personal grounds for behaving in an environmentally friendly way (obtain reward or avoid punishment), sanctioned laws, financial incentives, and preferential treatment schemes send a strong signal of commitment by the authorities to solving the targeted problem. Besides often being conspicuous in themselves, these types of measures tend to get a lot of publicity. Hence, they may be instrumental in increasing public attention to, and reasoning about, the targeted problem. It has been suggested that in some cases they may even facilitate the internalisation of behavioural norms relating to the problem. For example, Kahle and Beatty (1987) ascribed the higher environmental awareness of Oregonian youth compared to youth from other American states to the former's "growing up with the bottle bill", the (in an American context) innovative deposit-law regarding beverage containers that Oregon implemented one and a half decades earlier.[25]

---

25. One may doubt whether an effect like this can be produced by the incentive of a container deposit. However, if it is true, as implied by Kahle and Beatty, that Oregon's deposit law became a symbol of innovative environmental protection, the claim that it has raised environmental awareness seems more justified.

Further, if restraint is hampered by low trust in others doing their share (which is, of course, usually a prerequisite for the environmental problem being solved), coercive means are generally more effective trust-builders than non-coercive means (Messick, Wilke, Brewer, Kramer, Zemke, & Lui, 1983; Samuelson, 1993; Samuelson, Messick, Rutte, & Wilke, 1984; Yamagishi, 1986). For example, a recent Danish survey[26] found that 46% of the respondents preferred voluntary measures to increase recycling of household waste, but only 20% believed that such measures would be the most effective. In fact, 33% believed that financial incentives, and 44% that legal regulations would be the most effective way to increase recycling, indicating a lack of trust that others would recycle without some form of coercion.[27]

Negative side effects. Coercive means are likely to be perceived as infringing on individual freedom (Samuelson, 1993; Samuelson & Messick, 1995). Psychological reactance theory suggests that a perceived threat to one's freedom results in a motivational state directed at engaging in the threatened free behaviour (Brehm, 1966; Brehm & Brehm, 1981; Clee & Wicklund, 1980). Psychological reactance is particularly likely if the individual expects a free choice, and if the freedom is important to him or her.

The literature reports several examples of environmental regulation dependent on coercion being undermined or weakened due to psychological reactance (e.g., Gardner & Stern, 1996; Mazis, 1975; Van Vugt, Van Lange, Meertens, & Joireman, 1996). In a study of the effects of a ban on phosphate detergents in Miami, Florida, Mazis (1975) found that, after the ban, a sample of Miami consumers rated phosphate detergents more favourably and anti-pollution policy less favourably than a matched sample of citizens from another metropolitan area in Florida (Tampa) where no phosphate law had been passed. Similarly, a before/after field study of the effects of a Dutch carpool-priority lane on solo drivers' judgments and preferences regarding commuting alone or

26. A CATI based survey based on a random sample of 1000 respondents made in the summer of 1995 by the first author of this chapter with Suzanne C. Beckmann, Erik Kloppenborg Madsen, and Folke Ölander. The survey was financed by the Danish Government's Strategic Environmental Research Programme as a part of these scholars' research projects in CeSaM.

27. However, lack of trust did not seem to be a serious impediment to recycling. For example, 86% of the respondents claimed that they recycled all of their glass waste, and an additional 9% claimed that they recycled some of it.

by carpool found that solo drivers' preferences for carpooling tended to decline rather than to increase after the establishment of the carpool lane (Van Vugt et al., 1996). Rather than adapting their behaviour, solo drivers tended to adapt their beliefs in a way that justified continuing to drive alone (by upgrading the importance of an attribute strongly linked to driving alone and by downgrading the importance of an attribute strongly linked to carpooling). Among the visual signs of reactance were people driving alone in the carpool lane; some with mannequins in their cars. In the phosphate study, 12 % of the Miami sample reported buying phosphate detergents from neighbouring counties (Mazis, Settle, & Leslie, 1973).

Influential writers in the history of social dilemma research have argued that distributional inequality is a prerequisite for saving scarce resources (Edney, 1980). However, recent research has found that fear of new terms leading to greater inequality reduces the acceptance of stricter means of regulating access (Samuelson, 1993; Samuelson & Messick, 1995). Experimental research indicates that the reaction of individuals facing new constraints depends not only on the change in their own situation, but also on equality and equity considerations (Kerr, 1995). The same has been found in field research. For example, a review of energy conservation literature concludes that "if the 'sacrifices' required by an energy conservation measure are seen as affecting various population segments equally, the measures will be more readily accepted" (McClelland & Canter, 1981, p. 13). Similarly, a study of volume- and weight- based garbage handling fees in two rural municipalities in Denmark found that the attitude towards this regulatory measure was significantly influenced by its perceived fairness (Thøgersen, 1994b). About a third of a random sample of citizens from the two municipalities felt that it was unfair. The perceived fairness of the measure was uncorrelated with household size and expected personal gain, making equity concerns the most likely source of reservations.

A final negative side effect of coercive regulation, early suggested by Stern and Kirkpatrick (1977, p. 12), is that it has "the potential of converting people who value ... conservation into people who conserve only because it pays". The suggestion is supported by experimental research showing that a positive attitude towards an activity can be undermined by "overjustification" (De Young & Kaplan, 1985-86; Katzev & Pardini, 1987), when people are given an extrinsic reason for doing something they would have done anyway. For example, in a field experiment conducted in two areas in Switzerland designated by the Swiss government as potential sites for a nuclear waste repository facility, Frey and Oberholzer-Gee (1997) found that offering a monetary compensation for hosting such a facility in one's community drastically reduced public support. When the monetary compensation was offered, an indicator of perceived civic duty to accept the facility ceased to influence the decision on whether or not to accept it.

Research like this has led to the suggestion that the initiation, and especially the continuation, of wanted behaviour are more effectively produced by small than by large inducements (Festinger & Carlsmith, 1959). The effectiveness of the "minimal justification" principle in motivating environmentally beneficial behaviour has been demonstrated in several field experiments (Katzev & Johnson, 1984; Katzev & Pardini, 1987). Wiener and Doescher (1991), on the other hand, argue that a small personal incentive may also reduce restraint: a small personal incentive may be enough to increase the importance of the private payoff, but if the incentive is small, the sacrifices involved may easily be perceived to be greater than the advantages (including the incentive). For example, it has been found that the introduction of a small monetary incentive to promote blood donation actually decreased the quantity supplied (Frey, 1993).

Deci and Ryan (1985), and later Frey (1993), have suggested that whether or not a reward reduces the receiver's intrinsic motivation depends not on its size, but on whether or not it is contingent on his or her task engagement and performance. When it obviously is, receivers as well as observers tend to attribute pro-social action to the reward. When it is not (i.e., when rewards are given as tokens of appreciation, like medals, words of honour, food and drink served in connection with voluntary work, etc.), pro-social action is attributed to the actors' intrinsic motivation.

Non-coercive means

Positive side effects. The positive and negative side effects of non-coercive means are more or less the reverse of those of coercive means. When restraint is evoked by non-coercive means, people are more likely to attribute their restraint to internalised attitudes and norms (Festinger & Carlsmith, 1959; Freedman & Fraser, 1966; Katzev & Johnson, 1983; Katzev & Pardini, 1987). A commitment to show restraint created under these circumstances is likely to be stronger, more durable, and to have a stronger and more consistent influence on behaviour than a commitment produced by coercive means (Arbuthnot, Tedeschi, Wayner, Turner, Kressel, & Rush, 1977; Katzev & Johnson, 1983; Katzev & Johnson, 1984; Katzev & Pardini, 1987; Pallak, Cook, & Sullivan, 1980). Further, non-coercive means can evoke a sense of participation and personal responsibility for solving the problem that can rarely be obtained by coercive means. Among other advantages, a high sense of participation may prompt the targets of the regulation to generate valuable suggestions for improving its effectiveness (Gardner & Stern, 1996). In addition, when internalised attitudes and norms for environmentally friendly conduct are made salient (which is more likely to happen as the result of regulation by non-coercive than by coercive means) they may "spill over" and promote environmental friendliness in other fields than that targeted by the regulation (Thøgersen, 1999). For example, recent studies indicate that people who perform environmentally friendly activities on a voluntary basis are more likely than others to perform other environmentally friendly activities (Berger, 1997; Thøgersen, 1999; Thøgersen & Ölander, 1999).

Negative side effects. The most important negative side effect of using non-coercive means (especially information) is that it may convey the impression that the issue has low priority, and that the authorities are not inclined to do anything effective to solve the problem (Breck, 1997; Johansson, 1993). Hence, relying on such means alone may reduce trust in the problem ever being solved and increase non-restraint, due to fear of being a sucker (Wiener & Doescher, 1991).

SUMMARY, PERSPECTIVES, AND IMPLICATIONS

The conflict between individual consumption and environmental protection

Many, if not all, consumption-related environmental problems involve a conflict between individual and collective goals. Cars provide convenient, fast means of transportation, but they emit polluting substances, annoying noise, and threats of physical harm, and when travel by car exceeds a certain threshold, long since passed, these emissions cause serious environmental damage. PVC packaging is cheap and provides consumers with many desirable benefits, but it eventually enters the waste stream, and if incinerated it adds dioxins and other serious pollutants to the emissions from incinerator smokestacks, threatening the health of people living or consuming products that are grown "downwind." Modern, chemically-based farming puts cheap food products on supermarket shelves, but pesticides pollute aquifers at fearsome speed and over-fertilization upsets the ecological balance in inland and coastal waters (with catastrophic consequences for biotopes and fishery).

In addition to involving conflicts between individual and collective goals, solving environmental problems is complicated by the fact that their causes, and especially the relationship of these causes to individual behaviour, are not easily comprehensible to the layperson. In fact, these causes and relationships are often perceived as uncertain even by the scientific community, as is obvious in the current debate about global warming. Consumers and others who are willing to choose what seems to be the collectively preferable option in spite of their individual self-interest are typically acting on hunches or simple heuristics rather than on unambiguous knowledge.

Environmental problems as N-person games

Most consumption-related environmental problems are structurally similar to the *N*-person game (Dawes, 1980). The air, the sea, an underground aquifer, a lake, a species of fish, a scenic landscape, and many other natural phenomena have in common the fact that they serve as resources with – in principle – free access to (large groups of) people to use as they want, including as waste receptors.[28] As long as the total use of such a resource does not exceed its replenishing (or self-cleansing) capacity, there is no environmental problem (at least from the point of view of humans). But if resource use exceeds this threshold, the resource deteriorates. Fish yields fall, the beauty of the landscape vanishes, water from an aquifer is no longer potable, etc.

Even if they know that the resource deteriorates because of overuse, individual users adhering to the axioms of standard rational choice theory are unlikely to restrain their use of the resource voluntarily. While each individual's resource use (typically) negligibly contributes to the deterioration of the resource, it gives the individual substantial benefits. And giving up (part of) these benefits by restraining one's use of the

28. The resources mentioned are renewable (or self-cleaning) within limits. Other natural resources, like oil, gas, and minerals, are non-renewable. The case of non-renewable is not substantially different from that of renewable resources with regard to the matters discussed here (see, e.g., Daly, 1990).

resource only means that someone else reaps the benefits. Hence, it is individually rational to continue to use as much as one wants or can until the resource is no longer available; thus leaving everyone who depended on the resource worse off than if everyone had shown restraint.

A number of deviations from rationality have been mentioned. Some of them reduce the likelihood that individuals will show restraint. Notably, environmental problems are often characterized by a high degree of uncertainty. Often costs and benefits of environmentally sensitive behaviours are unequally distributed in time, producing what Platt (1973) called a "social trap." The pollution of streams and underground aquifers with pesticides from agriculture and gardening, for example, is typically not noticed until years after the farmers and gardeners began to reap the benefits of spraying. (Often it takes even more time and large sums of money to clean up the mess. In the worst cases, the environmental harm is irreversible, as when a species has become extinct.) Often the size of the resource is not known with certainty, leaving decision-makers prey to wishful thinking. Due to uncertainty and perceptual biases, people may not realize that the threshold has been exceeded until the resource is doomed to extinction.

Other deviations from rationality increase the likelihood of restraint. Under conditions that facilitate trust in a sufficient number of others doing the same and faith in the resource being saved, most people seem to show restraint; contrary to the predictions of standard rational choice theory. Strong social or internalised norms dictating restraint may explain this. A recent study by Thøgersen (1998) in a Danish context finds that everyday consumer activities, such as recycling and packaging waste, and to a lesser degree buying organic food, are guided by internalised norms.

Implications

The most radical way of solving many environmental problems is to change their structural characteristics so that they no longer pose a social dilemma, i.e., so that individual and collective rational behaviour converge. However, in many important cases this is simply not possible, due either to lack of scientific knowledge or control possibilities, or because such a solution would create conflicts with other important goals like democracy or equity. Hence, regulation authorities are often restricted to trying to reduce the size of the social dilemma and/or to trying to convince individuals to show restraint in spite of the dilemma.

In cases where the regulation is based on the assumption that people are rational, in the sense of standard rational choice theory, lack of realism may weaken the regulation. Regulations acknowledging humans' limited mental capacity, lack of knowledge, and uncertainty about future events stand much higher chances of success.

However, some frequently reported deviations from standard rational choice theory increase the possibilities of solving environmental problems. It is especially noteworthy that behaviour in social dilemmas is often based on norms rather than on deliberate calculations of "what's in it for me." Norm-based decision-making may have been adaptive in the development of our species, both because it increases the likelihood of survival of individuals living in groups (Simon, 1990) and because it demands less cognitive effort than a utility-maximizing decision (Payne et al., 1990).

In order for a regulatory means to influence behaviour, it must be apprehended by the target population. Hence, the effects of a regulation depend on how people perceive the regulation. The research reviewed shows that the important thing is not only how a target individual perceives that the regulation influences his or her own conditions, but also how he or she perceives that it influences the conditions of others, the nature of the problem, and the authorities' commitment to solve it.

Basically, regulation should target the most important barriers to restraint, or the "limiting factors", to use an expression from ecology (Gardner & Stern, 1996). For instance, it is often found that people are highly motivated to recycle, and that the most important barrier to recycling is lack of convenient recycling opportunities (e.g., Berger, 1997; Guagnano, Stern, & Dietz, 1995; Thøgersen, 1994a). Under these circumstances, the recent fashion of adding an economic incentive scheme, thus providing an additional motive for recycling, is hardly helpful (Ackerman, 1997; Thøgersen, 1994b).

With luck, supplying external motives to people who are already intrinsically motivated to show restraint has no effect (apart from waste of money and effort by the authorities). In less fortunate circumstances, external motives may displace or "crowd out" (Frey, 1993; Frey & Oberholzer-Gee, 1997) intrinsic motives, possibly leaving the targeted individuals less motivated, particularly in the longer run.

REFERENCES

Ackerman, F. (1997). *Why do we recycle? Markets, values, and public policy.* Washington, D.C.: Island Press.

Arbuthnot, J., Tedeschi, R., Wayner, M., Turner, J., Kressel, S. & Rush, R. (1977). The induction of sustained recycling behavior through the foot-in-the-door technique. *J. Environmental Systems, 6*(4), 355-369.

Axelrod, R. (1984). *The evolution of cooperation.* New York: Basic Books.

Bateson, C. D., Dyck, L., Brandt, J. R., Bateson, J. G., Powell, A. L., McMaster, M. R. & Griffitt, C. (1988). Five studies testing two new egoistic alternatives to the empathy-altruism hypothesis. *Journal of Personality and Social Psychology, 55*, 52-77.

Berger, I. E. (1997). The demographics of recycling and the structure of environmental behavior. *Environment and Behavior, 29*, 515-531.

Biel, A. & Gärling, T. (1995). The role of uncertainty in resource dilemmas. *Journal of Environmental Psychology, 15*, 221-233.

Breck, T. (1997). Forbrugerpolitik eller miljøpolitik? *Samfundsøkonomen,* (2), 37-41.

Brehm, J. W. (1966). *A theory of psychological reactance.* New York: Academic Press.

Brehm, S. S. & Brehm, J. W. (1981). *Psychological reactance: A theory of freedom and control.* San Diego: Academic Press.

Brennan, G. (1990). What might rationality fail to do? In: Cook, K. S. & Levi, M. (Ed.). *The limits to rationality*, pp. 51-59. Chicago: The University of Chicago Press.

Bruins, J. J., Liebrand, W. B. G. & Wilke, H. A. M. (1989). About the saliency of fear and greed in social dilemmas. *European Journal of Social Psychology, 19*, 155-161.

Budescu, D. V. & Bruderman, M. (1995). The relationship between the illusion of control and desirability bias. *Journal of Behavioral Decision Making, 8*, 109-125.

Budescu, D. V., Rapoport, A. & Suleiman, R. (1990). Resource dilemmas with environmental uncertainty and asymmetric players. *European Journal of Social Psychology, 20*, 475-487.

Camerer, C. (1995). Individual decision making. In: Kagel, J. H. & Roth, A. E. (Ed.). *The handbook of experimental economics*, pp. 587-703. Princeton: Princeton University Press.

Camerer, C. F. & Johnson, E. J. (1991). The process-performance paradox in expert judgment. In: Ericson, K. A. & Smith, J. eds.). *Toward a general theory of expertise,* pp. 195-217. Cambridge: Cambridge University Press.

Caporael, L., Dawes, R. M., Orbell, J. M. & Kragt, A. J. C. v. d. (1989). Selfishness examined: Cooperation in the absence of egoistic incentives. *Behavioral and Brain Sciences, 12,* 683-739.

Clee, M. & Wicklund, R. (1980). Consumer behavior and psychological reactance. *Journal of Consumer Research, 6,* 389-405.

Daly, H. (1990). Towards some operational principles of sustainable development. *Ecological Economics, 2,* 1-6.

Dawes, R. M. (1980). Social dilemmas. *Annual Review of Psychology, 31,* 169-193.

Dawes, R. M. (1998). Behavioral decision making, judgment, and inference. In: Gilbert, D., Fiske, S. & Lindzey, G. eds.). *The handbook of social psychology (3rd ed.),* pp. 497-548. Boston: McGraw-Hill.

De Young, R. & Kaplan, S. (1985-86). Conservation behavior and the structure of satisfactions. *Journal of Environmental Systems, 15*(3), 223-292.

Dietz, T. & Stern, P. C. (1995). Toward a theory of choice: Socially embedded preference construction. *Journal of Socio-Economics, 24,* 261-279.

Edney, J. (1980). The commons problem: Alternative perspectives. *American Psychologist, 35,* 131-150.

Etzioni, A. (1988). *The moral dimension. Toward a new economics.* New York: The Free Press.

Festinger, L. & Carlsmith, J. M. (1959). Cognitive consequences of forced compliance. *Journal of Abnormal and Social Psychology, 58,* 203-210.

Ford, J. K., Schmitt, N., Schechtman, S. L., Hults, B. M. & Doherty, M. L. (1989). Process tracing methods: Problems and neglected research questions. *Organizational Behavior and Human Decision Processes, 43,* 75-117.

Freedman, J. & Fraser, S. (1966). Compliance without pressure: The foot-in-the-door technique. *Journal of Personality and Social Psychology, 4,* 195-202.

Frey, B. S. (1993). Motivation as a limit to pricing. *Journal of Economic Psychology, 14,* 635-664.

Frey, B. & Oberholzer-Gee, F. (1997). The cost of price incentives: An empirical analysis of motivation crowding-out. *American Economic Review, 87,* 746-755.

Frisch, D. & Clemen, R. T. (1994). Beyond expected utility: Rethinking behavioral decision research. *Psychological Bulletin, 116*(no 1), 46-54.

Gardner, G. T. & Stern, P. C. (1996). *Environmental problems and human behavior.* Boston: Allyn and Bacon.

Gärling, T. (1999). Value priorities, social value orientations, and cooperation in social dilemmas. *British Journal of Social Psychology, 38,* 397-408.

Gärling, T., Karlsson, N., Romanus, J. & Selart, M. (1997). Influences of the past on choices of the future. In: Ranyard, R., Crozier, R. & Svensson (Ed.). *Decision making: Models and explanations.* London: Routledge.

Gärling, T., Gustafsson, M., & Biel, A. (1999). Managing uncertain common resources. In M. Foddy, M. Smithson, M. Hogg, & S. Schneider (Eds.), *Resolving social dilemmas* (pp. 219-225). Philadelphia, PA: Psychology Press.

Guagnano, G. A., Stern, P. C. & Dietz, T. (1995). Influences on attitude-behavior relationships. A natural experiment with curbside recycling. *Environment and Behavior, 27,* 699-718.

Hardin, G. (1968). Tragedy of the commons. *Science, 162*(13), 1243-1248.

Hogarth, R. M. & Einhorn, H. J. (1990). Venture theory: A model of decision weights. *Management Science, 36,* 780-803.

Johansson, M. G. (1993). *Et adfærdsteoretisk grundlag for regulering af husholdningernes affaldsbortskaffelse.* Aarhus: Institut for Markedsøkonomi, Handelshøjskolen i Aarhus. ("Regulation of households' waste disposal, based on behavourial theory". The Dept. of Marketing, The Århus School of Business)

Kahle, L. R. & Beatty, S. E. (1987). Cognitive consequences of legislating postpurchase behavior: growing up with the bottle bill. *Journal of Applied Social Psychology, 17*(9), 828-843.

Kahneman, D. (1994). New challenges to the rationality assumption. *Journal of Instrumental and Theoretical Economics (JITE), 150*(1), 18-36.

Kahneman, D., Slovic, P. & Tversky, A. (Ed.) (1982). *Judgement under uncertainty: Heuristics and biases.* Cambridge: Cambridge University Press.

Kahneman, D. & Tversky, A. (1979). Prospect theory: An analysis of decision making under risk. *Econometrica, 47,* 263-291.

Katzev, R. D. & Johnson, T. R. (1983). A social-psychological analysis of residential electricity consumption: the impact of minimal justification techniques. *Journal of Economic Psychology*, (3), 267-284.

Katzev, R. D. & Johnson, T. R. (1984). Comparing the effects of monetary incentives and foot-in-the-door strategies in promoting residential energy conservation. *Journal of Applied Social Psychology, 14*, 12-27.

Katzev, R. D. & Pardini, A. U. (1987). The comparative effectiveness of reward and commitment approaches in motivating community recycling. *Journal of Environmental Systems, 17*(2), 93-113.

Kerr, N. L. (1995). Norms in Social Dilemmas. In: Schroeder, D. A. (Ed.). *Social Dilemmas. Perspectives on Individuals and Groups* (1.), pp. 31-48. Westport, Connecticut: Praeger Pulishers.

Kerr, N. L. & Kaufman-Gilliland, C. M. (1994). Communication, commitment, and cooperation in social dilemmas. *Journal of Personality and Social Psychology, 66*, 513-529.

Klandermans, B. (1992). Persuasive communication: Measures to overcome real-life social dilemmas. In: Liebrand, W., Messick, D. M. & Wilke, H. (Ed.). *Social dilemmas: Theoretical issues and research findings*, pp. 307-318. Oxford: Pergamon Press.

Komorita, S. S. (1976). A model of the *N*-person dilemma-type game. *Journal of Experimental Social Psychology, 12*, 357-373.

Komorita, S. S. & Barth, J. M. (1985). Components of reward in social dilemmas. *Journal of Personality and Social Psychology, 48*, 364-373.

Kulhman, D. M. & Marshello, A. (1975). Individual differences in game motivation as moderators of preprogrammed strategic effects in prisoner's dilemma. *Journal of Personality and Social Psychology, 46*, 1044-1057.

Levi, M., Cook, K. S., O'Brien, J. A. & Faye, H. (1990). Introduction: The limits of rationality. In: Cook, K. S. & Levi, M. (Ed.). *The limits of rationality*, pp. 1-16. Chicago: The University of Chicago Press.

Liebrand, W. B. G. & McClintoch, C. (1988). The ring measure of social values: A computerized procedure for assessing individual differences in information processing and social value orientation. *European Journal of Personality, 2*, 217-230.

Lopes, L. L. (1987). Between hope and fear: The psychology of risk. *Advances in Experimental Psychology, 20*, 255-295.

Luce, R. D. & Raiffa, H. (1957). *Games and decisions: Introduction and critical survey*. New York: John Wiley and Sons.

Mansbridge, J. J. (1990). *Beyond self-interest*. Chicago: University of Chicago Press.

Mazis, M. B. (1975). Antipollution measures and psychological reactance theory: A field experiment. *Journal of Personality and Social Psychology, 31*(4), 654-660.

Mazis, M. B., Settle, R. B. & Leslie, D. C. (1973). Elimination of phosphate detergents and psychological reactance. *Journal of Marketing Research, 10*, 390-395.

McClelland, L. & Canter, R. J. (1981). Psychological research on energy conservation: Context, approaches, methods. In: Baum, A. & Singer, J. E. (Ed.). *Advances in environmental psychology, Vol. 3. Energy conservation: Psychological perspectives*, pp. 1-26. Hillsdale: Lawrence Erlbaum.

Messick, D. M. & Brewer, M. B. (1983). Solving social dilemmas: A review. In: Wheeler, L. & Shaver, P. (Ed.). *Review of personality and social psychology*, pp. 11-44. Beverly Hills: Sage.

Messick, D. M. & McClelland, C. L. (1983). Social traps and temporal traps. *Personality and Social Psychology Bulletin, 9*, 105-110.

Messick, D. M., Wilke, H. A. M., Brewer, M. B., Kramer, R. M., Zemke, P. E. & Lui, L. (1983). Individual adaptations and structural change as solutions to social dilemmas. *Journal of Personality and Social Psychology, 44*, 294-309.

Olson, M. (1965). *The logic of collective action*. Cambridge: Harvard University Press.

Ophuls, W. (1973). Leviathan or oblivion? In: Daly, H. E. (Ed.). *Toward a steady state economy*, pp. 215-230. San Francisco: Freeman.

Orbell, J. M., Kragt, A. J. C. v. d. & Dawes, R. M. (1988). Explaining discussion-induced cooperation. *Journal of Personality and Social Psychology, 54*, 811-819.

Ostrom, E. (1990). *Governing the commons. The evolution of institutions for collective action*. Cambridge: Cambridge University Press.

Pallak, M. S., Cook, D. A. & Sullivan, J. J. (1980). Commitment and energy conservation. In: Bickman, L. (Ed.). *Applied social psychology annual, Vol. 1*, pp. 235-253. Beverly Hills: Sage.

Payne, J. W., Bettman, J. R. & Johnson, E. J. (1988). Adaptive strategy selection in decision making. *Journal of Experimental Psychology: Learning, Memory, and Cognition, 14*, 534-552.

Payne, J. W., Bettman, J. R. & Johnson, E. J. (1990). The adaptive decision maker: Effort and accuracy in choice. In: Hogarth, R. M. (Ed.). *Insights in decision making: A tribute to Hillel J. Einhorn*, pp. 129-153. Chicago: University of Chicago Press.

Platt, J. (1973). Social traps. *American Psychologist, 28*, 641-651.

Pruitt, D. G. & Kimmle, M. (1977). Twenty years of experimental gaming: Critique, synthesis, and suggestions for the future. *Annual Review of Psychology, 28*, 363-392.

Rapoport, A., Budescu, D. V., Suleiman, R. & Weg, E. (1992). Social dilemmas with uniformly distributed resources. In: Liebrand, W., Messick, D. M. & Wilke, H. (Ed.). *Social dilemmas: Theoretical issues and research findings*, pp. 43-57. Oxford: Pergamon Press.

Samuelson, C. D. (1993). A multivariate evaluation approach to structural change in resource dilemmas. *Organizational Behavior and Human Decision Processes, 55*, 298-324.

Samuelson, C. D. & Messick, D. M. (1995). When Do People Want to Change the Rules for Allocating Shared Resources? In: Schroeder, D. A. (Ed.). *Social Dilemmas. Perspectives on Individuals and Groups* (1.), pp. 143-162. Westport, Connecticut: Praeger Pulishers.

Samuelson, C. D., Messick, D. M., Rutte, C. G. & Wilke, H. A. M. (1984). Individual and structural solutions to resource dilemmas in two cultures. *Journal of Personality and Social Psychology, 44*, 294-309.

Savage, L. J. (1954). *The foundation of statistics*. New York: Wiley.

Schwartz, S. H. (1977). Normative influence on altruism. In: Berkowitz, L. (ed. *Advances in experimental social psychology, Vol. 10*, pp. 221-279. New York: Academic Press.

Schwartz, S. H. (1992). Universals in the content and structure of values: Theoretical advances and empirical tests in 20 countries. In: Zanna, M. P. (Ed.). *Advances in Experimental Social Psychology, Vol. 25*, pp. 1-65. San Diego: Academic Press.

Schwartz, S. H. (1994). Are there universal aspects in the structure and content of human values? *Journal of Social Issues, 50*(4), 19-45.

Schwartz, S. H. & Bilsky, W. (1987). Toward a universal psychological structure of human values. *Journal of Personality and Social Psychology, 53*, 550-562.

Schwartz, S. H. & Bilsky, W. (1990). Toward a theory of universal content and structure of values: Extensions and cross-cultural replications. *Journal of Personality and Social Psychology, 58*(5), 878-891.

Schwartz, S. H. & Sagiv, L. (1995). Identifying culture-specifics in the content and structure of values. *Journal of Cross-Cultural Psychology, 26*(1), 92-116.

Sen, A. (1990). Rational behaviour. In: Eatwell, J., Milgate, M. & Newman, P. (Ed.). *The new palgrave: Utility and probability*. New York: W.W. Norton & Company.

Simon, H. (1990). A mechanism for social selection and successful altruism. *Science, 250*, 1665-1668.

Simon, H. A. (1982). *Models of bounded rationality. Vol. 2: Behavioral economics and business organization*. Cambridge: The MIT Press.

Stern, P. C. & Kirkpatrick, E. M. (1977). Energy behavior: Conservation without coercion. *Environment, 10*, 10-15.

Thøgersen, J. (1994a). A model of recycling behaviour. With evidence from Danish source separation programmes. *International Journal of Research in Marketing, 11*, 145-163.

Thøgersen, J. (1994b). Monetary incentives and environmental concern. Effects of a differentiated garbage fee. *Journal of Consumer Policy, 17*, 407-442.

Thøgersen, J. (1998). *The norm-attitude-behavior relationship. Theory and applications in the environmental domain*. Working Paper (98-3). Aarhus: Aarhus School of Business.

Thøgersen, J. (1999). Spillover processes in the development of a sustainable consumption pattern. *Journal of Economic Psychology, 20*, 53-81.

Thøgersen, J. & Ölander, F. (1999). Danske forbrugeres oplevelse af forskelle og ligheder mellem miljørelevante aktiviteter. (Danish consumers' perception of differences and similarities among environmentally relevant activities). *Working Paper 1999-3*. Århus: Marketing og Miljø Forskningsgruppen, Institut for Markedsøkonomi, Handelshøjskolen i Århus. (Marketing and environment research group, The Århus School of Business)

Tversky, A. & Fox, C. R. (1995). Weighing risk and uncertainty. *Psychological Review, 102*, 269-283.

Tversky, A. & Kahneman, D. (1986). Rational choice and the framing of decisions. *Journal of Business, 59*(4 (pt. 2)), 251-278.

Tversky, A. & Shafir, E. (1992). The disjunction effect in choice under uncertainty. *Psychological Science, 3*, 305-309.

Van Dijk, E. & Wilke, E. (1995). Coordination rules in asymmetric social dilemmas: A comparison between public good dilemmas and resource dilemmas. *Journal of Experimental Social Psychology, 31*, 1-27.

Van Lange, P. A. M., Liebrand, W. B. G., Messick, D. M. & Wilke, H. A. M. (1992). Introduction and literature review. In: Liebrand, W. B. G., Messick, D. M. & Wilke, H. A. M. (Ed.). *Social dilemmas: Theoretical issues and research findings*, pp. 43-57. Oxford: Pergamon Press.

Van Vugt, M., Van Lange, P. A. M., Meertens, R. M. & Joireman, J. A. (1996). How a structural solution to a real-world social dilemma failed: A field experiment on the first carpool lane in Europe. *Social Psychology Quarterly, 59*, 364-374.

Von Neumann, J. & Morgenstern, O. (1944). *Theory of games and economic behavior.* Princeton: Princeton University Press.

Wiener, J. L. & Doescher, T. A. (1991). A framework for promoting cooperation. *Journal of Marketing,* (April), 38-47.

Wilke, H. A. M. (1991). Greed, efficiency, and fairness in resource management situations. In: Stroebe, W. & Hewstone, M. (Ed.). *European Review of Social Psychology, Vol. 2,* John Wiley & Sons Ltd.

Yamagishi, T. (1986). The structural goal/expectation theory of cooperation in social dilemmas. In: Lawler, E. & Markovsky, B. (Ed.). *Advances in group processes, Vol. 3*, pp. 51-87. Greenwich: JAI Press.

Yamagishi, T. (1988). The provision of a sanction system in the United States and Japan. *Social Psychology Quarterly, 51*, 264-270.

Zaltman, G. (1974). Strategies for diffusing innovations. In: Sheth, J. N. & Wright, P. L. (Ed.). *Marketing Analysis for Societal Problems*, pp. 78-100. Urbana-Champaign: The University of Illinois.

About the Contributors

SVEND ANDERSEN

Professor of ethics and philosophy of religion at the Faculty of Theology, University of Aarhus, and director of the Centre for Bioethics, University of Aarhus. He gained a doctorate in theology from the University of Heidelberg (thesis on the philosophy of Kant), and a doctorate in theology from the University of Aarhus (thesis on the philosophy of religious language). Among his books are: *Ideal und Singularität. Über die Funktion des Gottesbegriffes in Kants theoretischer Philosophie* (1983) and *Einführung in die Ethik* (2000). Member of the Danish Council of Ethics from 1988-1993, and guest professor at the University of Kiel in 1994. Since 1999 he has been president of Societas Ethica, European Society on Research in Ethics.

HELLE TEGNER ANKER

Associate Professor, Ph.D. (1996). Legal scholar and researcher at CeSaM/School of Law, University of Aarhus. Board member of CeSaM. Her main fields of research are general environmental law issues at international and national levels. Specific areas of research are biodiversity, nature protection, integrated coastal zone management, agriculture and other land use activities. Author of 'Miljøretlig regulering på landbrugsområdet' [Environmental Regulation Within the Agricultural Sector] (1996), and co-editor of 'Land Use and Nature Protection – Emerging Legal Aspects' (2000).

ELLEN MARGRETHE BASSE

Professor, *Dr.jur. h.c.* Since 1995 legal professor of environmental law at the School of Law, University of Aarhus, and from 1992 director of the Centre for Social Science Research on the Environment (CeSaM). In 2000 appointed doctor *honoris causa*. Main fields of research are general environmental issues at international, EU and national level. Specific areas of research are environmental management, standardization, certification, classification, environmental agreement and other market-based instruments as part of the norm-based strategies. She has published more than 90 books, her latest work being *Miljøretten* – 6 volumes providing an overall insight into the many rules, principles and other regulatory instruments within environment and trade.

SUZANNE C. BECKMANN

Dr. *rer.soc.* Strategic Planning Director at Saatchi & Saatchi and Professor at the Department of Intercultural Communication and Management, Copenhagen Business School. Visiting Professor at the Pennsylvania State University, at the University of Utah, the University of Innsbruck, Aston Business School in Birmingham, and the University of Melbourne. Member of the editorial board of the *International Journal of Research in Marketing*, the *Journal of Economic Psychology*, the *Journal of Consumer Policy* and *Virksomhedens miljøhåndbog*. Research interests include consumer behaviour, communication, and strategic management. Her research has been widely published in international journals and anthologies.

MARTIN ENEVOLDSEN

Assistant Professor in public policy at the Department of Political Science, University of Aarhus. Most of his research and publications are within the subject areas of environmental policy and energy policy. Lecturer in environmental economics at The Royal Veterinary and Agricultural University in Copenhagen. Presently completing his Ph.D. thesis 'Environmental agreements and taxes: a new institutional approach to effective CO_2 regulation' which is an attempt to establish a rational institutionalist approach to study the effectiveness of indirect environmental regulation.

TOMMY GÄRLING

Graduate of Stockholm University, Ph.D. in psychology. Formerly Associate Professor at Umeå University and Professor in Psychology and Planning at the Swedish

Research Council for the Humanities and Social Sciences. Since 1996, Professor of Psychology, Göteborg University. Co-editor of six books, and a forthcoming title *Residential Choice and Satisfaction* soon to be published by Greenwood. Has authored and co-authored more than 200 articles and reports. Current research includes the role of actual and anticipated emotional states in decision making, household financial management, and municipality decision making on, for example, road pricing.

WILLIAM E. KILBOURNE

Gained his Ph.D. in marketing from the University of Houston, 1973. Has taught at the University of Houston, California State University, Humboldt, Oklahoma State University, and is currently at Sam Houston State University in Texas. He has also held visiting positions at the University of Colorado, Denver and Copenhagen Business School. His research has appeared in the *Journal of Marketing Research*, the *Journal of Advertising*, the *Journal of the Academy of Marketing Science*, the *Journal of Business Research*, the *Journal of Macromarketing*, and the *Journal of Marketing Management*. Current research interest is in the relationships between business, consumption, and the natural environment.

ERIK KLOPPENBORG MADSEN

Tenured member of the Faculty of Business Administration at the Aarhus School of Business from which he received his Ph.D. in business economics in 1989. As Associate Professor at the Department of Marketing he has taught, besides general marketing, philosophy of science and ethics for business students and his research and writings have focussed on and aired criticism concerning the dominant concept of instrumental rationality in economics and business studies.

FOLKE ÖLANDER

Ph.D., *Dr. h.c.*, Professor of Economic Psychology at the Aarhus School of Business. His interest in the environmental impact of consumers' behaviour is of fairly recent origin. He has published several books and journal articles over the past 30 years. Currently editor of the *Journal of Consumer Policy*, and past President of the International Association for Research in Economic Psychology (IAREP). Former member and chairman, Danish Social Science Research Council. In 1999, appointed Chairman of an Advisory Committee to the Danish Minister of Trade and Industry; member of the Nordic Consumer Committee. Held visiting professorships in Canada, Norway and New Zealand.

JOHN THØGERSEN

Ph.D., *Dr. merc.* Professor of Economic Psychology at the Aarhus School of Business. Has published extensively on environmental attitudes and behaviour issues, some of which have appeared in journals such as, the *Journal of Economic Psychology*, the *Journal of Consumer Policy*, *Psychology & Marketing*, *Environment & Behavior*, *International Journal of Research in Marketing*, and *Business Strategy and the Environment*. He is the coordinator of the Business and Environment Research Theme at the Faculty of Business Administration, Aarhus School of Business, and member of the board of the Centre for Transport Research on Environmental and Health Impacts and Policy, TRIP. Former member of the board of CeSaM, and member of the Danish Technology Assessment Council's expert committee on recycling.

ULLI ZEITLER

Graduated from the University of Aarhus with a degree in philosophy and history in 1984. Has taught political philosophy, ethics and philosophy of science at the Universities of Aarhus and Copenhagen and the Aarhus Business School. From 1994 to 1999 was a full-time researcher at CeSaM where he received his Ph.D. in philosophy and held post-doctoral positions in transport and environmental ethics. Assistant Professor in philosophy of social sciences and political philosophy at Aalborg University from 1999 to 2001. Currently project manager for Aarhus County Administration, Department of Education and Labour Market. Has published books on philosophy of production and transport ethics, and various articles within the field of environmental ethics.